Electric Circuits

TEACHER'S GUIDE

SCIENCE AND TECHNOLOGY FOR CHILDREN

NATIONAL SCIENCE RESOURCES CENTER
Smithsonian Institution–National Academy of Sciences
Arts and Industries Building, Room 1201
Washington, DC 20560

NSRC

The National Science Resources Center is operated by the National Academy of Sciences and the Smithsonian Institution to improve the teaching of science in the nation's schools. The NSRC collects and disseminates information about exemplary teaching resources, develops and disseminates curriculum materials, and sponsors outreach activities, specifically in the areas of leadership development and technical assistance, to help school districts develop and sustain hands-on science programs. The NSRC is located in the Arts and Industries Building of the Smithsonian Institution and in the Capital Gallery Building in Washington, D.C.

National Science Resources Center

Douglas Lapp, Executive Director
Charles N. Hardy, Deputy Director for Information Dissemination, Materials Development, and Publications
Sally Goetz Shuler, Deputy Director for Development, External Relations, and Outreach
R. Gail Thomas, Administrative Officer
Anne E. Pomerleau, Financial Associate
Terence Proctor, Information/Technology Specialist
Gail Greenberg, Executive Administrative Assistant
Katherine Darke, Administrative Assistant
Tonya Miller, Receptionist/Office Assistant
Kathleen Holmay, Public Information Consultant
Leslie O'Flahaven, Development Consultant

Publications
Dean Trackman, Publications Director
Lynn Miller, Writer/Editor
Max-Karl Winkler, Illustrator
Heidi Kupke, Publications Technology Specialist
David Stein, Editorial Assistant
Laura Akgulian, Writer/Editor Consultant
Cynthia Allen, Writer/Editor Consultant
Judith Grumstrup-Scott, Writer/Editor Consultant
Linda Harteker, Writer/Editor Consultant
Dorothy Sawicki, Writer/Editor Consultant
Lois Sloan, Illustrator Consultant

Science and Technology for Children
Joyce Lowry Weiskopf, Director
Wendy Binder, Research Associate
Edward V. Lee, Research Associate
Christopher Lyon, Research Associate
Amanda Revere, Office Assistant
Don Cammiso, Research Consultant
Carol O'Donnell, Research Consultant
Judith White, Research Consultant

Outreach
Julie Clyman Lee, Program Associate
Cathy Gruber, Program Assistant
L. J. Benton, Consultant

Information Dissemination
Evelyn M. Ernst, Director
Marilyn Fenichel, Program Officer
Ted Schultz, Program Officer
Rita C. Warpeha, Resource/Database Specialist
Barbara K. Johnson, Research Associate
Sharon S. Seaward, Program Assistant

The above individuals were members of the NSRC staff in 1995.

Copyright © 1991, National Academy of Sciences. All rights reserved.
99 98 97 96 95   10 9 8 7 6 5

ISBN 0-89278-661-2

Published by Carolina Biological Supply Company, 2700 York Road, Burlington, NC 27215.
Call toll free 1-800-334-5551.

No part of this book may be reproduced by any mechanical, photographic, or electronic process, or in the form of a phonographic recording, nor may it be stored in a retrieval system, transmitted, or otherwise copied for public or private use without permission in writing from the National Science Resources Center.

See specific instructions in lessons and appendices for photocopying.

♻ Printed on recycled paper.

Foreword

Why study science? Science, more than any other discipline, provides us with tools to learn about the world. Science is not a listing of facts; science is an invitation to observe the world, ask questions, and puzzle over problems and enjoy the process of solving them. From the time children begin to perceive their environment, they are involved in science. This kind of science is the core of the NSRC's Science and Technology for Children (STC) project; its goal is to provide teachers and children with an inquiry-based curriculum that builds on children's excitement as they discover new things and guides them in arriving at explanations that make sense to them.

This unit is one of 24 elementary science curriculum units being developed by the STC project for grades one through six. Each STC unit provides children with the opportunity to learn in-depth about topics in the physical, life, or earth sciences and technology through direct observation and experimentation. The units invite children to develop hypotheses, then to test their ideas, just as professional scientists do. Discovering what makes a light bulb turn on and the way plants grow are just as exciting—and important—events to young children as the discoveries that scientists make. Along the way, children develop patience, persistence, and confidence in their own ability to tackle and solve real problems.

As the teacher, your role in this process is crucial. You will be guiding hands-on learning—encouraging students to explore new ideas for themselves. This is a rewarding way of teaching—and exciting. You will be helping students learn to think for themselves as they expand their understanding of the world around them.

Acknowledgments

The National Science Resources Center would like to acknowledge the many individuals and organizations that have supported the Science and Technology for Children Project. The major funding for the project is from the John D. and Catherine T. MacArthur Foundation, with additional funds from the Dow Chemical Company Foundation; E.I. du Pont de Nemours and Company; Amoco Foundation, Inc.; the Hewlett-Packard Company; the U.S. Department of Defense; and the U.S. Department of Education.

The primary author of the *Electric Circuits* unit was Joe Griffith. In developing this unit, he worked closely with a number of schools, including the Watkins School of the Capitol Hill Cluster Schools in Washington, D.C., where he trial-taught the unit. He also worked with a number of individuals, including:

L.J. Benton, Coordinator, Instructional Materials Processing Center, Fairfax County Public Schools, Fairfax, VA

Cary Burkey, Teacher, Watkins School

Suzanne Clack, Teacher, Reynolds School District, Portland, OR

JoAnn DeMaria, Teacher, Hutchison Elementary School, Herndon, VA

Peggy Freund, Teacher, Markham Elementary School, Fort Belvoir, VA

Jack Fry, Teacher, Paschal Sherman Indian School, Omak, WA

Ann Gay, Vice Principal, Capitol Hill Cluster Schools

Joy Goode, Teacher, Carole Highlands School, Takoma Park, MD

Veola Jackson, Principal, Watkins School;

Ulysses Johnson, Vice Principal, Watkins School

Michael Jordan, Project Director for Electricity Exhibit, Science Museum of Virginia, Richmond, VA

Dick McQueen, Science Specialist, Multnomah Education Service District, Portland, OR

Perry Montoya, Teacher, Johnson Elementary School, Mesa, AZ

Dane Penland, Staff Photographer, Smithsonian Institution Photographic Services

Jerome Pine, Professor of Physics, California Institute of Technology, Pasadena, CA

Gennis Powell, Vice Principal, Watkins School

Barbara Scott, Principal, Carole Highlands School, Takoma Park, MD

Catherine Taylor, Elementary Science Resource Teacher, Watkins School

The NSRC would like to thank George Hein and Sabra Price of Lesley College for evaluating this unit. The NSRC also would like to thank the following individuals and school systems for their assistance with the national field testing of the *Electric Circuits* unit:

Judi Backman, Math/Science Coordinator, Highline Public Schools, Seattle, WA

Tom Campbell, Adams Elementary School, Midland, MI

Diane Cote, Chippewassee Elementary School, Midland, MI

Nancy Graham, Adams Elementary School, Midland, MI

Ray Hart, Olympic School, Seattle, WA

Teri Johnston, Hybla Valley School, Fairfax County Public Schools, Fairfax, VA

Sarah Lindsey, Science Coordinator, Midland Public Schools, Midland, MI

Carolyn Preston, Bunker Hill School, Washington, DC

Gayle Richards, Department of Defense Dependent Schools, Clark Air Base, The Philippines

Jan Smith, Midway Elementary School, Des Moines, WA

Joyce Turner, Bunker Hill School, Washington, DC

Mary Whitley, Hilltop Elementary School, Seattle, WA

We also are indebted to the members of the STC Advisory Panel, listed below, who reviewed the unit and made suggestions for its improvement.

We would like to thank Robert McC. Adams, Secretary of the Smithsonian Institution; Frank Press, President of the National Academy of Sciences; and the NSRC Advisory Board for their vision and support in helping the NSRC to undertake this project.

Douglas Lapp
Executive Director
National Science Resources Center

STC Advisory Panel

Peter Afflerbach, Associate Professor, National Reading Research Center, University of Maryland at College Park

David Babcock, Director, Board of Cooperative Educational Services, Second Supervisory District, Monroe-Orleans Counties, Spencerport, New York

Judi Backman, Math/Science Coordinator, Highline Public Schools, Seattle, Washington

Albert Baez, President, Vivamos Mejor/USA, Greenbrae, California

Andrew R. Barron, Professor of Chemistry, Harvard University, Cambridge, Massachusetts

DeAnna Banks Beane, Project Director, YouthALIVE, Association of Science-Technology Centers, Washington, D.C.

Al Buccino, Education Advisor, Office of Science and Technology Policy, Executive Office of the White House, Washington, D.C.

Audrey Champagne, Professor of Chemistry and Education, and Chair, Educational Theory and Practice, School of Education, State University of New York at Albany, Albany, New York

Sally Crissman, Faculty Member, Lower School, Shady Hill School, Cambridge, Massachusetts

Gregory Crosby, National Program Leader, U.S. Department of Agriculture Extension Service/4-H, Washington, D.C.

JoAnn E. DeMaria, Teacher, Hutchison Elementary School, Herndon, Virginia

Hubert M. Dyasi, Director, Workshop Center for Open Education, City College of New York, New York, New York

Timothy H. Goldsmith, Professor of Biology, Yale University, New Haven, Connecticut

Charles N. Hardy, Assistant Superintendent, Instruction and Curriculum, Highline Public Schools, Seattle, Washington

Patricia Jacobberger Jellison, Geologist, National Air and Space Museum, Smithsonian Institution, Washington, D.C.

Patricia Lauber, Author, Weston, Connecticut

John Layman, Professor of Physics, University of Maryland, College Park, Maryland

Sally Love, Museum Specialist, National Museum of Natural History, Smithsonian Institution, Washington, D.C.

Phyllis R. Marcuccio, Assistant Executive Director of Publications, National Science Teachers Association, Arlington, Virginia

Lynn Margulis, Professor of Biology, University of Massachusetts, Amherst, Massachusetts

Margo A. Mastropieri, Co-Director, Mainstreaming Handicapped Students in Science Project, Purdue University, West Lafayette, Indiana

Richard McQueen, Specialist, Science Education, Multnomah Education Service District, Portland, Oregon

Alan Mehler, Professor, Department of Biochemistry and Molecular Science, College of Medicine, Howard University, Washington, D.C.

Philip Morrison, Professor of Physics, Emeritus, Massachusetts Institute of Technology, Cambridge, Massachusetts

Phylis Morrison, Educational Consultant, Cambridge, Massachusetts

Fran Nankin, Editor, *SuperScience Red*, Scholastic, Inc., New York, New York

Jerome Pine, Professor of Physics, California Institute of Technology, Pasadena, California

Harold Pratt, Director, Middle School Science Project, Jefferson County Public Schools, Golden, Colorado

Wayne E. Ransom, Executive Director of Educational Programs, Franklin Institute, Philadelphia, Pennsylvania

David Reuther, Editor-in-Chief and Vice-President, William Morrow Books, New York, New York

Robert Ridky, Associate Professor of Geology, University of Maryland, College Park, Maryland

F. James Rutherford, Chief Education Officer and Director, Project 2061, American Association for the Advancement of Science, Washington, D.C.

David Savage, Training Specialist, Office of Instruction and Program Development,

Montgomery County Public Schools, Rockville, Maryland

Thomas E. Scruggs, Co-Director, Mainstreaming Handicapped Students in Science Project, Purdue University, West Lafayette, Indiana

Larry Small, Science/Health Coordinator, Schaumburg School District 54, Schaumburg, Illinois

Michelle Smith, Publications Coordinator, Office of Elementary and Secondary Education, Smithsonian Institution, Washington, D.C.

Susan Sprague, Director of Science and Social Studies, Mesa Public Schools, Mesa, Arizona

Arthur Sussman, Director, Far West Regional Consortium for Science and Mathematics, Far West Laboratory, San Francisco, California

Emma Walton, Science Program Coordinator, Anchorage School District, and Past President, National Science Supervisors Association, Anchorage, Alaska

Paul Williams, Director, Center for Biology Education, and Professor, Department of Plant Pathology, University of Wisconsin, Madison, Wisconsin

Kathryn Wolff, Managing Editor, American Association for the Advancement of Science, Washington, D.C.

Contents

	Foreword	iii
	Acknowledgments	iv
	Contents	vii
	Unit Overview and Materials List	1
	Teaching Strategies and Classroom Management Tips	3
Lesson 1	Thinking about Electricity and Its Properties	7
Lesson 2	What Electricity Can Do	11
Lesson 3	A Closer Look at Circuits	19
Lesson 4	What Is Inside a Light Bulb?	25
Lesson 5	Building a Circuit	29
Lesson 6	What's Wrong with the Circuit?	37
Lesson 7	Conductors and Insulators	43
Lesson 8	Making a Filament	49
Lesson 9	Hidden Circuits	53
Lesson 10	Deciphering a Secret Language	57
Lesson 11	Exploring Series and Parallel Circuits	63
Lesson 12	Learning about Switches	69
Lesson 13	Constructing a Flashlight	73
Lesson 14	Working with a Diode	77
Lesson 15	Planning to Wire a House	81
Lesson 16	Wiring and Lighting the House	85
Appendix A	Post-Unit Assessments	89
Appendix B	Teacher's Record Chart of Student Progress	97
Appendix C	Using the Cutting and Stripping Tool	99
Appendix D	Removing the Base From a Light Bulb	103
Appendix E	Making and Troubleshooting Hidden Circuit Boxes	107
Appendix F	Background for Lesson 11: Series and Parallel Circuits	109
Appendix G	Bibliography	113
Appendix H	Materials Reorder Information	117

Goals for *Electric Circuits*

In this unit, students expand their understanding of electricity through investigations with wires, batteries, bulbs, and switches. Their experiences introduce them to the following concepts, skills, and attitudes.

Concepts

- A complete electric circuit is required for electricity to light a bulb.
- A complete circuit can be constructed in more than one way using the same materials.
- Different types of electric circuits show different characteristics.
- A switch can be used to complete or interrupt a circuit.
- Some materials conduct electricity; these are called conductors.
- Some materials do not conduct electricity; these are called insulators.
- Electricity can produce light and heat.
- A diode conducts electricity in one direction only.

Skills

- Wiring simple electric circuits.
- Predicting, observing, describing, and recording results of experiments with electricity.
- Drawing conclusions about circuits from the results of experiments.
- Building and using a simple circuit tester.
- Using symbols to represent the different parts of an electric circuit.
- Building a simple switch.
- Applying troubleshooting strategies to complete an incomplete circuit.
- Applying information about electric circuits to design and build a flashlight.
- Applying information about electric circuits to design and wire a house.
- Reading to learn more about electricity.
- Communicating results and ideas through writing, drawing, and discussion.

Attitudes

- Appreciating the need for safety rules when working with electricity.
- Developing an interest in electricity.
- Developing confidence in being able to analyze and solve a problem.

Unit Overview and Materials List

The modern world would not be possible without electricity. Electricity lights our homes and industries; powers many forms of transportation; supplies complex lines of communication through telephones, televisions, radios, and computers; and provides us with a dizzying array of labor-saving devices, both in the home and at the workplace. What's more, a knowledge of electricity has given scientists new ways to pursue other disciplines—chemistry, physics, biology, and medicine.

Children, too, are fascinated with electricity. They wonder how it powers so many different kinds of devices. This unit, *Electric Circuits*, will help them begin to answer their many questions by opening the door to the world of electricity. The unit makes use of electrical "stuff," such as wires, bulbs, and batteries, and includes investigations that are both useful and fun. Although the unit was originally designed and tested for grade 4, it could be taught at grades 3 or 5, as well.

Electric Circuits is divided into three parts. In the first part, Lessons 1 through 6, students are introduced to the basic properties of electricity and learn about electric circuits and the parts of a light bulb. During the middle section, Lessons 7 through 10, students learn about conductors and insulators. They also learn about the symbols used to represent the parts of a circuit in circuit diagrams. In the last half of the unit, Lessons 11 through 16, students explore different kinds of circuits, learn about switches, construct a flashlight, and discover the properties of diodes. The unit culminates with students wiring a cardboard box house.

The **Appendices** include many suggestions for post-unit assessments. They also provide information about preparing materials and constructing hidden circuits boxes. The **Bibliography** has suggestions of books that will help students see how discoveries in electricity fit into the history of the United States, as well as books that will help students delve further into the scientific and technological aspects of electricity.

You do not have to be an expert in electricity to teach this unit. The background sections of the Teacher's Guide will provide you with most of the information you need. But don't be surprised if you find yourself learning along with the students, and if you and your students find yourselves faced with puzzling questions. Use this situation to model the way scientists learn: define the question, then ask, "How can we find out?" This will encourage your students to find out on their own by experimenting and consulting resource materials.

Materials List

Below is a list of materials needed for the *Electric Circuits* unit.

- 1 Teacher's Guide
- 15 Student Activity Books
- 60 D-cell batteries
- 60 No. 48 bulbs (microlamp)
- 2 household bulbs, 60 watt, clear glass
- 1 roll (100 feet) #22 coated hook-up wire
- 1 roll (15 feet) #32 nichrome wire
- 30 battery holders
- 30 bulb sockets
- 15 diodes, 1N4007
- 15 packages of assorted objects, each containing:
    - golf tee
    - 1-inch piece of soda straw
    - brass screw
    - paper clip
    - aluminum screening (1-inch square)
    - plastic screening (1-inch square)
    - 1-inch piece of chalk
    - wooden pencil stub (no eraser, lead exposed at both ends)
    - brass paper fastener
    - wire nail
    - aluminum nail
    - marble
    - 1-inch piece of pipe cleaner
    - 1-inch piece of bare copper wire
    - 1-inch piece of bare aluminum wire
- 200 Fahnestock clips
- 1 wire stripper tool
- 1 wire cutter
- 1 pair of needle-nose pliers
- 2 screwdrivers
- 45 small boxes (2" x 4" x 7½") for storage and hidden circuit boxes
- 2 boxes of No. 1 paper clips (100)
- 2 boxes of No. 3 brass paper fasteners
- 2 boxes of washers for brass paper fasteners
- 1 pound of modeling clay
- 100 3" x 5" cards
- 30 labels
- 8 rolls masking tape
- 1 faulty No. 48 bulb
- * Student notebooks
- * Glue stick
- * 24" x 36" newsprint pad
- * Overhead projector and markers
- *8 pairs of scissors
- *1 sheet fine sandpaper
- * Crayons or markers
- * Pencils
- * Cardboard boxes
- * 8½" x 11" sheets of drawing paper
- * 8½" x 11" sheets of construction paper

*Note: These items are not included in the kit. Including them would increase material and shipping costs, and they are commonly available in most schools or can be brought from home.

Teaching Strategies and Classroom Management Tips

The teaching strategies and classroom management tips in this section will help you give students the guidance they need to make the most of their hands-on experiences in this unit. These strategies and tips are based on the understanding that students already have knowledge and ideas about how the world works. And that useful learning results when they have the opportunity to think about their ideas as they engage in new experiences and encounter the ideas of others.

Classroom Discussion: Discussions effectively led by the teacher are important. Research shows that the way questions are asked as well as the time allowed for responses can contribute to the quality of the discussion. When you ask questions, think about what you want to achieve in the ensuing discussion. For example, open-ended questions, for which there is clearly no one right answer, will encourage students to give creative and thoughtful answers. Other types of questions can be used to encourage students to see specific relationships and contrasts or to help students to summarize and draw conclusions. It is good practice to mix these questions. It also is good practice to always give the students "wait-time" to answer; that time (some researchers recommend a minimum of 3 seconds) will buy you broader participation and more thoughtful answers.

Brainstorming: A brainstorming session is a whole-class exercise in which students contribute their thoughts about a particular idea or problem. It can be a stimulating and productive exercise when used to introduce a new science topic. It is also a useful and efficient way for the teacher to find out what students know and think about a topic. As students learn the rules for brainstorming, they will become more and more adept in their participation.

To begin a session, define for students the topics about which ideas will be shared. Tell students the following rules:

- Accept all ideas without judgment.
- Don't criticize or make unnecessary comments about the contributions of others.
- Try to hitch your ideas onto the ideas of others.

Ways to Group Students: One of the best ways to teach hands-on science lessons is to arrange students in small groups of two to four. There are several advantages to this organization. It offers students a chance to learn from one another by sharing ideas, discoveries, and skills. With coaching, students can develop important interpersonal skills that will serve them well in all aspects of life. Finally, by having students in groups assist each other, you will have more time to work with those students who need the most help.

As students work, often it will be productive for them to talk about what they are doing, resulting in a steady hum of conversation. If you or others in the school are accustomed to a completely quiet room at all times, this new, busy atmosphere may require some adjustment. It will be important, of course, to establish some limits to keep the noise under control.

Safety: One objective of this unit is to teach students how to have a healthy respect for the hazards of electric power. Students also will learn how to perform some safe experiments with electricity. Below are some specific safety instructions:

- Caution your class not to experiment with electrical outlets and appliances at home or in school. The high voltage supplied from these outlets can deliver a fatal shock.

- The batteries used in this unit will not give a shock unless more than two dozen are connected in series. Even then, the batteries will produce only a mild shock. Household electricity, on the other hand, can deliver a lethal shock.

- Do not use rechargeable batteries. There have been reports of very hot wires when these batteries are short-circuited.

- If a light bulb should break, there will probably be broken glass on the floor. Establish a procedure to clean up the glass. Ask students to let you know if there is any broken glass so you can supervise cleanup.

Handling Materials: There will be student materials and projects to be stored from lesson to lesson. A good system for storing and dispensing materials will save time and tempers. For student storage, standard-sized folding boxes (called "folding mailers") work well because they are sturdy and stack neatly. By using stick-on labels for student names, the boxes can be recycled for use with another class. Shoe boxes are another alternative.

A complete list of materials needed for this unit is included at the end of the **Unit Overview**. When using hook-up wire, you will need to strip about ¾ of an inch from each end so it can be used in circuits. (For instructions about stripping the wire, see **Appendix C.**)

Frequently, extra supplies from the teacher's box will need to be dispensed. As the unit unfolds, students will need replacement batteries or bulbs and extra wire. By controlling the supply box, the teacher maintains steady contact with student activities, but at times dispensing these supplies will get hectic. Another adult in the room to take charge of supplies will free the teacher to interact with the students so there are fewer distractions. Another alternative would be to have students take the responsibility for dispensing these additional supplies on a rotating basis as one of several housekeeping tasks.

Setting Up a Learning Center: Students can benefit greatly from having some supplies available for them to work with in their spare time. One way to accomplish this is to arrange a table with a set or two of the materials for students to use. Many lessons include information about materials that might be put in a center, and questions that students can explore.

Curriculum Integration: There are many opportunities for curriculum integration in this unit. Look for the following icons for math, reading, writing, art, and speaking that highlight these opportunities.

Evaluation: Evaluation suggestions are included at regular intervals throughout the unit. These suggestions should help the teacher assess what students know and monitor how they are progressing. With that information it is possible for the teacher to provide assistance to students who are struggling.

Appendix A offers post-unit assessments that can be used to find out what students learned in this unit. A selection is offered so that the teacher can choose the most appropriate assessments for the students. If a pre-test is desirable, brainstorming lists from Lesson 1 and the student drawing of a bulb from the beginning of Lesson 2 could suffice. A matching post-test could then include student descriptions of what they have learned and their final drawings of the light bulb.

Appendix B provides a black line master for a "Teacher's Record Chart of Student Progress" that you may choose to use as a checklist. It could help you in your recording for the class and in your identification of students who are not keeping up. Please keep in mind that most fourth graders will not be able to master and articulate this full list of skills.

Portfolios of student work also are useful in many ways, for example, sharing with parents or other interested adults. A portfolio could include student notebooks with drawings, writings, diagrams, completed activity sheets, and descriptions of projects. Other pieces of student work, such as the specific circuits they create, the flashlight they design, and the house they wire provide concrete demonstrations of student knowledge. Finally, student presentations can be useful vehicles for assessment as well as useful language and critical thinking experiences in which students formulate and articulate their ideas.

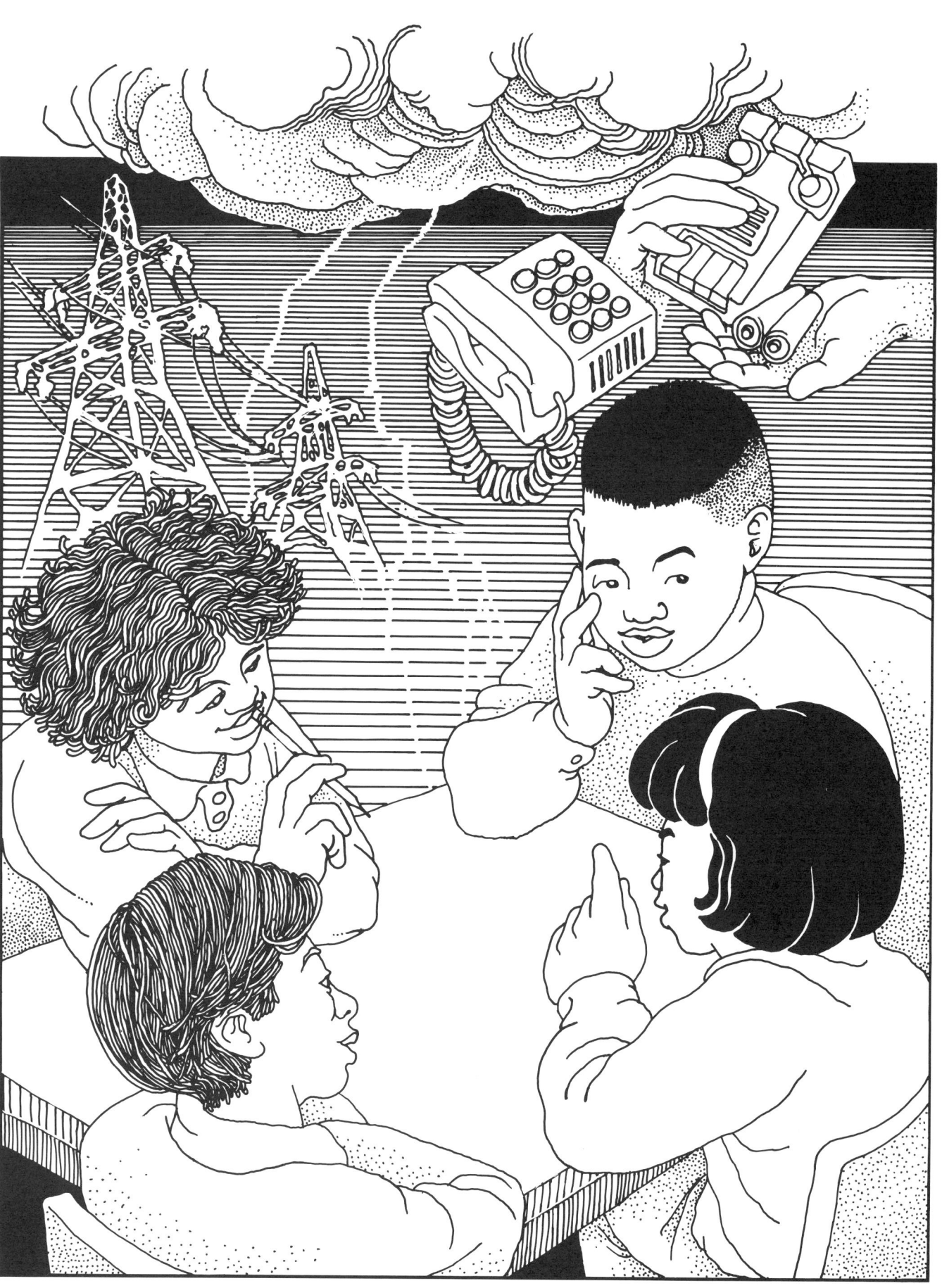

LESSON 1

Thinking about Electricity and Its Properties

Overview

In this lesson, students think about what they know about electricity and what they would like to learn. As the discussion develops, students probably will be surprised at how much they already know about the uses of electricity. By the end of the lesson, they will have a better sense of what they would like to know about how electricity works.

Objectives

- Students learn brainstorming techniques.
- Students discuss what they know about electricity and what they would like to learn.
- Students review important safety rules about using electricity.

Background

Just a mere 200 years ago, electricity had not yet been discovered. Life was very different back then. People lit their homes with candles or whale oil. They sent messages by word of mouth and letters by foot, horse, or boat.

In 1752, Benjamin Franklin paved the way for an understanding of electricity with his famous experiments. He revealed that lightning was electricity. In the early 1800s, Michael Faraday added to this knowledge by discovering the relationship between magnetism and electricity. In the mid-1800s, Joseph Henry revealed the nature of electromagnetic induction. The foundation was then laid for a score of inventions, from the light bulb by Thomas Edison to the telegraph by Samuel Morse. Today, we continue to add to the list of inventions that depend on electricity.

For this lesson, keep in mind that brainstorming sessions can be very stimulating and productive exercises when you are introducing a new science topic. As students learn the rules for these sessions, they will become more and more adept at participating in them.

See **Teaching Strategies and Classroom Management Tips** for more information about brainstorming.

LESSON 1

Materials

For each student
1 pencil
1 student notebook

For the class
To record contributions, use one of the following:
1 24" x 36" newsprint pad and markers
  Chalkboard and chalk
  Overhead transparencies and a marker

Procedure

1. Tell your students that during this lesson they will have a chance to share what they know about electricity and what they would like to learn. Tell them that you are going to begin the unit with a brainstorming exercise. Let them know that all contributions will be accepted and no one is to criticize the ideas of others. See **Teaching Strategies and Classroom Management Tips** for more information about brainstorming.

   Here is one possible scenario:

   ■ Ask your students: "What do you know about electricity?" Record their responses on the chalkboard or on a piece of newsprint. When a duplicate response is given, put a check beside it to indicate that someone else has already thought of it. Keep a record of student responses to be used as part of your evaluation of the unit.

   ■ After about 10 minutes of brainstorming, while students are still excited, bring the session to a close by saying that you want to move on to another question.

   ■ Next, ask your students these questions: "Now that we have talked about what you know about electricity, let's think about a different idea. What questions do you have about electricity? What would you like to know?" Keep a record of student responses on a piece of newsprint.

   Note: As you teach the next 15 lessons, you might want to schedule some time to have your students add information to the lists about what they now know about electricity.

2. After you have completed your discussion about electricity, tell your students that there is another area you would like to address—working together. Tell them that from time to time we will talk about how we can work together well. Indicate that it is a challenge to learn to work with people—both the ones you like and the ones you don't like, and that at times, we'll work in pairs, in groups of four, and sometimes as a whole class, as well as alone.

3. Before going further in the unit, discuss with the class the difficulties they may have learning new skills. Tell the students that you want them to enjoy learning new things. Tell them that when we learn something new, we may be frustrated at first. But remind them that they can ask for help. When they give help to a classmate, they need to think about what kind of help is useful. Giving the answer is often not helpful. Usually it is better to give encouragement and hints. Most people learn best and enjoy it more by figuring something out for themselves, if they can.

4. Present the safety guidelines listed below to the class:

   ■ Do not experiment with the electricity in wall plugs, either in school or at home. Both of these currents can give a lethal shock and should not be used for experiments.

   Note: The batteries used in this unit will not give a noticeable shock until more than two dozen are connected in series, making their combined voltage about 36 volts. But even then a person would feel only a mild tingle.

   ■ Remember that electrical appliances that operate on household electricity can deliver shocks.

   ■ Avoid downed power lines and electrical substations. They can give deadly shocks.

5. Point out to the students that throughout the unit they will use their notebooks to write and draw many of their questions and observations.

Final Activities

End the lesson by reviewing the questions students have asked. Tell them that they will discover many of the answers to their questions by doing the experiments in the unit.

Extensions

1. Ask students to bring in lists or a collage of pictures illustrating all the ways they use electricity in their home. Share the lists by making a bulletin board or by having the students present their findings to the class.

2. Ask students to try to imagine what life would be like without electricity. What would they have to do for light? How would they accomplish chores such as washing clothes and cleaning the house? How else would life be different?

Evaluation

The brainstorming session will provide you with some information about each student's present ideas about electricity. It will provide a baseline against which to evaluate their progress as they work through the unit.

Thinking about Electricity and Its Properties / 9

| LESSON 2 | **What Electricity Can Do** |

Overview

Light bulbs are such an integral part of everyday life that most people can't imagine being without them. Because people tend to take light bulbs so much for granted, they don't think about how they work. But light bulbs are an ingenious electrical device. In this lesson, students will learn how to light a bulb using a simple battery, a piece of wire, and a small bulb. Students will be excited that they can get the bulb to light and will be interested in looking for different ways to complete the task.

Objectives

- Students discover how to light a bulb using a simple battery, a piece of wire, and a small bulb.
- Students set up a notebook for their observations.

Background

Electricity flows along a path called a circuit. To create a circuit, you need a battery, wire, and whatever else you wish to include in the circuit, such as a bulb. The electricity must be able to move from one end of the battery to the other to create a complete circuit.

How does electricity flow along a circuit? Like many things in nature, electricity is invisible, but we can see and measure the results of the flow. The battery, or energy source, gives electricity its "push" through a circuit. This push, or voltage, can be thought of as electrical pressure, and is analogous to water pressure. Electrical pressure is measured in volts.

The actual flow of electricity through a circuit is analogous to the flow of water through a hose. The flow of electrical current is measured in amperes.

Batteries come in many different shapes and sizes. High-voltage batteries are composed of cells, but the simplest batteries have only one cell. Common one-celled batteries are the AAA-, AA-, C-, and D-cells, all of which can be found in many local stores. Although these batteries differ in size and in the amount of current they can provide, they all produce approximately 1.5 volts.

In this unit, students will be working with a D-cell battery. The D-cell, like all batteries, has two ends, one marked + (positive) and one marked – (negative). The positive end has a small, raised button on it; the negative end is flat (see Figure 2-1).

LESSON 2

Materials can be roughly divided into two groups: conductors, which will readily transmit an electrical current; and insulators, which will not allow the flow of any significant amounts of electricity. Examples of conducting materials are aluminum, copper, and steel. Insulating materials include rubber, wood, and most plastics.

The wire students will be working with is commonly called "hook-up wire," because it is used to connect electrical components. The wire is made of copper or aluminum. The conducting wire allows electricity to flow from one end of the battery to the other. In some other wires, such as lamp cords, there are two wires, each composed of many strands. The two wires are necessary so that a complete circuit can be created, beginning at the power plant, going through a network of wires, to an appliance or light, and back to the power plant.

To prevent shocks, the wire is covered by a plastic sheath that acts as an insulator.

Figure 2-1

D-cell Batteries, hook-up wire and lamp cord

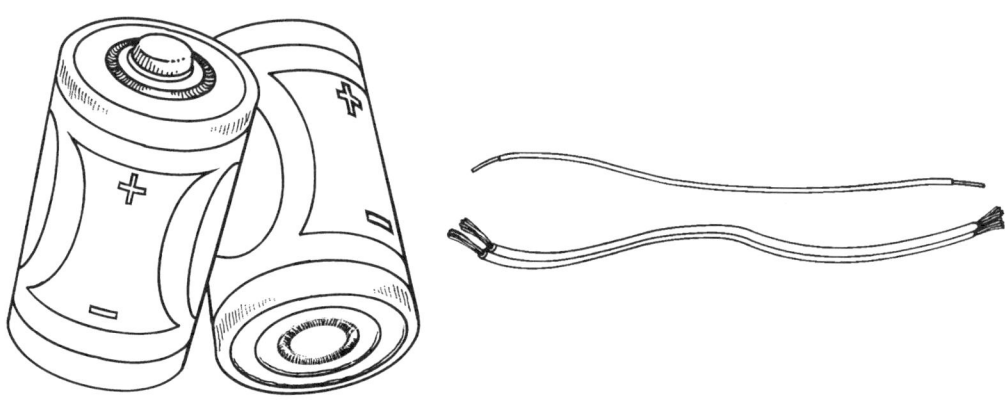

The bulbs used in this unit are very similar to the household bulbs in fixtures and lamps, except they are much smaller. A typical bulb is shown in Figure 2-2, and its parts have been labeled. The filament is the part of the bulb that gets hot, glows, and produces light.

Figure 2-2

A bulb

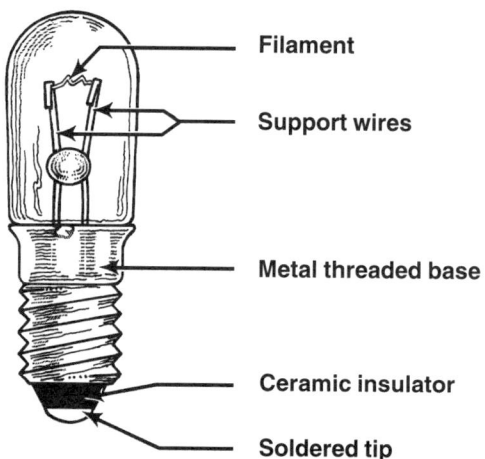

12 / What Electricity Can Do

All these components—the battery, the wire, and the bulb—can work together to make a circuit. The circuit is established when there is a continuous path for electricity to travel from one end of the D-cell back to the other end. Figure 2-3 shows a student lighting a bulb by using a piece of wire to connect the negative end of the D-cell to the bulb.

Figure 2-3

Lighting a bulb

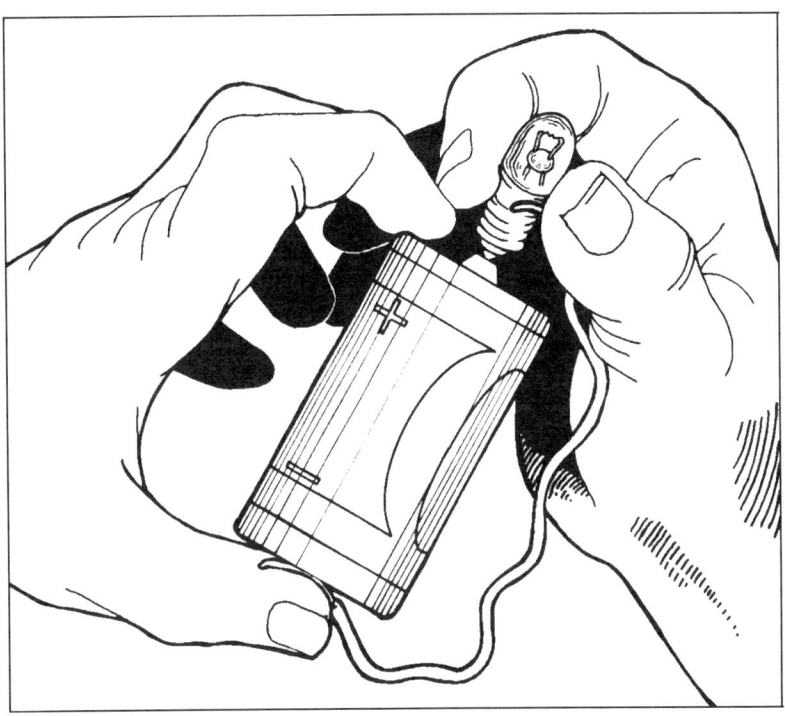

There are at least four different ways to light the bulb using one wire, one bulb, and one battery. These ways are illustrated in Figure 2-4. The bulb lights with the same brightness in each case.

Figure 2-4

Four ways to light a bulb

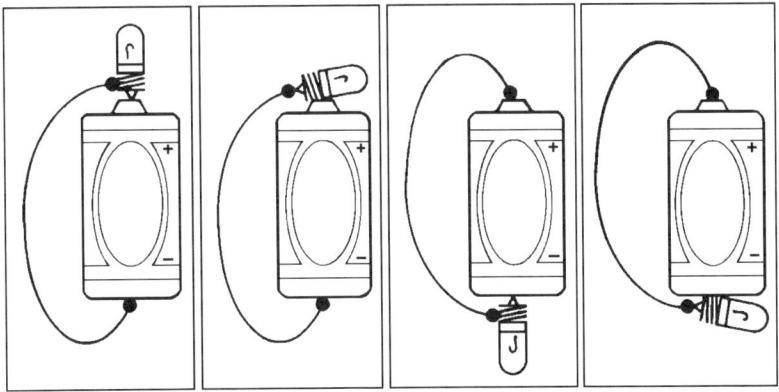

Sometimes students may connect a wire directly from one end of the battery to the other without having a bulb in the circuit. When this happens, a short circuit is created, as illustrated in Figure 2-5. A short circuit provides a conducting pathway from one end of the battery to the other, but bypasses the bulb.

Short circuits in a house or in an automobile can produce dramatic sparks and enough heat to melt metals and start a fire. Short circuits with D-cells

What Electricity Can Do / 13

Figure 2-5

A short circuit

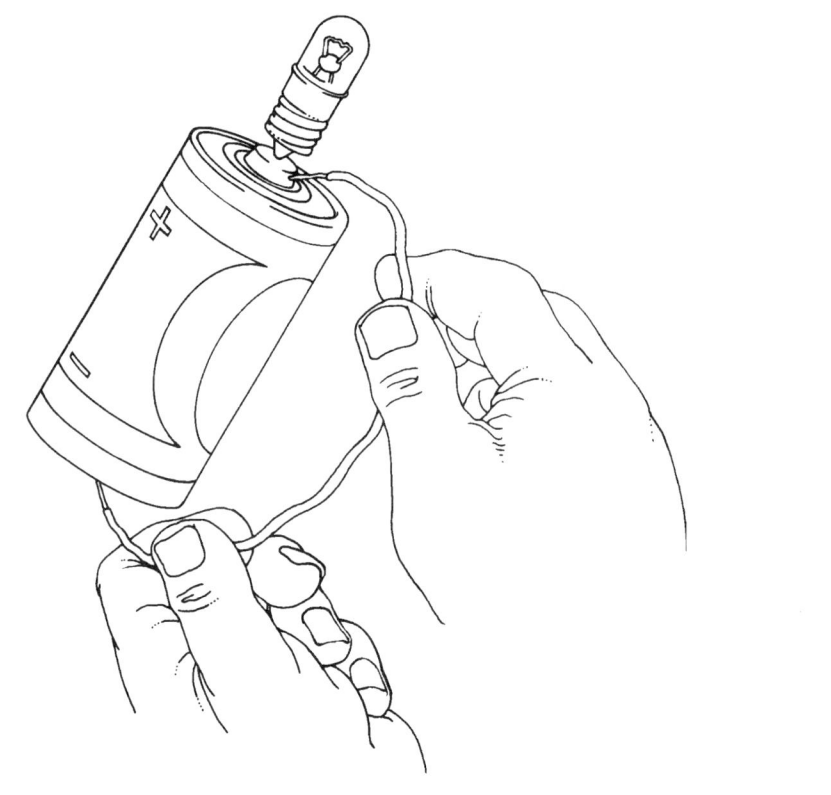

are not dangerous, but they do drain the electrical energy from the batteries. If left connected for several minutes, both the wire and D-cell will get warm.

As students work, they will inevitably create short circuits. Short circuits are not always easy to recognize, because they often occur amidst a tangle of fingers and wires. Two typical examples of the way such a short circuit might look are shown in Figure 2-6. A short circuit should be suspected when a student can't figure out why the bulb will not light. (Touching the sides of the D-cell will not cause a short circuit, however, nor will touching the D-cell with the insulated part of the wire.)

If a short circuit occurs but is disconnected immediately, the D-cell can continue to be used, as long as it still lights the bulb. However, if the D-cell will no longer light the bulb, it should be put aside. Save these for use in Lesson 6.

Figure 2-6

Two typical short circuits

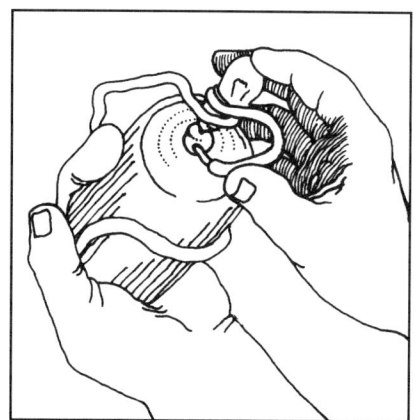 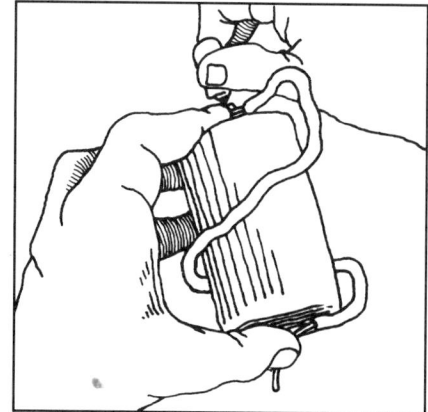

14 / What Electricity Can Do

It should be emphasized that there is no real risk in these short circuits and no damage is possible by students creating them. Do not demonstrate these as "wrong" things to do. This background is provided so you will be able to foresee simple stumbling blocks when they arise.

Materials

For each student

- 1 storage box
- 1 D-cell battery
- 1 bulb
- 1 6-inch piece of wire
- 1 label
- 1 student notebook

For the class

- 30 extra D-cell batteries
- 30 extra bulbs

(These are the reserves for the entire unit)

Safety Reminder

Do not use rechargeable batteries. There have been reports of very hot wires when these batteries are short-circuited.

Preparation
(1 hour)

1. Before the class, prepare the wire (instructions below) and assemble the student storage boxes, the D-cell batteries, and the bulbs. The wires will need to be stripped before you begin the unit. You may want to identify a few student helpers or an adult volunteer to work with you in stripping the ends of the wires and doing other assembly tasks.

2. To prepare the pieces of wire:
   - Cut one 6-inch piece of wire. To do this task, you can use a stripping tool like the one shown in **Appendix C**, or any similar device.
   - Next, strip approximately ¾ inch of insulation from each end of the wire. This is necessary so students can use the wire to connect their circuits.

3. Plan to have students gather their own set of materials.
   - Determine a location in the classroom where student access will be easy and where you can lay out the materials so the students can pick them up cafeteria style.
   - For each student, set out one storage box, one piece of 6-inch wire with the insulation stripped off each end, one D-cell, and one small bulb.

4. Take some time to create a circuit that will light the bulb. Before class, work through **Activity Sheet 1**, which is found at the end of Lesson 3. You will not use the sheet in this lesson, but it will help you to understand the different ways to light the bulb.

What Electricity Can Do / **15**

LESSON 2

Procedure

1. Ask your students to get out their notebooks. Remind students that they will be keeping a record of their findings and questions throughout the unit. The record will be writings as well as drawings.

   Now ask students to turn to the first page of their notebooks and to write the date on the page. Then have students draw a picture of a light bulb. Tell students to save this drawing; they will refer back to it later in the unit.

2. Tell the students to put their names on their labels, which they should then place on their boxes. Then the boxes can be reused by other classes.

3. Have the students assemble their set of materials by picking up one box, one wire, one D-cell battery, and one bulb. A typical distribution station is shown in Figure 2-7 below.

4. Ask students to open the boxes and remove the wire, the bulb, and the D-cell. Using only the wire, the bulb, and D-cell, ask each student to make the bulb light.

   Soon after students begin working, some will say:

   "I can't **do** this!"

   "Could you help me?"

   "This is **frustrating**!"

   As students work at this task, reassure them that it can be done. If they need coaching, ask: "How could you connect together the wires, the D-cell, and the bulb to get the bulb to light? What are some different ways you could do it?"

16 / What Electricity Can Do

The first student who succeeds in lighting the bulb usually lets out an excited yell. That initial excitement encourages other students to quickly solve the problem.

5. Have students draw pictures of the different ways the bulb lighted on one page of their notebook and of several ways it did not light on another page. Students should draw as many of their findings in their notebooks as they can during the lesson.

6. Be aware that some students will "short circuit" the D-cell by inadvertently connecting the wire from one end of the D-cell to the other, without having the electricity go through the bulb. Short circuits are shown in Figures 2-5 and 2-6 on pg. 14. When the D-cell or the wire become warm, look for a short circuit.

7. The initial excitement after first getting the bulb to light sometimes gives way to a desire to just see the bulb burn. Students may become very adept at finding small dark places in the room or in their desks that can be lit by the bulb. Allow children enough time to do these "experiments," if possible.

8. About ten minutes before the end of the period, have students clean up and put their materials away. This will leave time for a final discussion with the class. Have them check that only three items are in their boxes—the D-cell, the wire, and a bulb.

Final Activities

Ask the class to examine their drawings of the ways the bulb lighted and the ways it did not. Ask students to tell you something they have learned about how electricity works.

Extensions

It can be very useful at this point in the unit to set up a table with a box or two of the materials used in this lesson. This will give students a chance to experiment with circuits in their free time. Many students will benefit from this extra time working with the materials.

Evaluation

1. The drawings of the light bulbs will provide a useful contrast to the drawings the students make later in the unit. These drawings will provide a kind of pre- and post-test that can be used to show progress in the unit to parents or others who are interested.

2. Similarly, the students' drawings of the ways the bulb lights show how well they can observe intricate parts and how much they understand about electricity. They can also provide a baseline against which to compare later drawings.

3. Your observations of each student's approach to problem solving will provide useful information. Some students will systematically try all possibilities, while others will learn from their classmates. Still others will give up or feel defeated. These students will need encouragement.

LESSON 3	**A Closer Look at Circuits**

Overview

In Lesson 2, students learned that a complete circuit must be created in order for the bulb to light. In this lesson, students will reinforce that knowledge by discovering that there are different ways to create a complete circuit. This concept will be emphasized in future lessons.

Objectives

- Students review different ways to connect the battery, the wire, and the bulb to get the bulb to light.
- Students explore alternative ways to create a circuit.

Materials

For each student
- 1 storage box containing:
    - 1 D-cell battery
    - 1 bulb
    - 1 6-inch piece of wire
- 1 **Activity Sheet 1**
- 1 student notebook

For the class
- 10 copies of the large cut-out of a D-cell and a bulb (Figure 3-1)
- 1 glue stick
- 1 dark marker
- 1 24" x 36" newsprint pad

Preparation

Make several copies of the large, full-page drawing of the D-cell and bulb (Figure 3-1). Cut out the battery and the bulb pictures and paste them onto a large sheet of paper in various configurations in which the bulb could be lit using one wire. During the lesson, students will use a dark marker to draw wires illustrating ways they made the bulb light.

If you prefer to work on an overhead projector, use Figure 3-1 to make a transparency. Cut the battery and the bulb apart so they can be moved independently. Now the students can use a real piece of wire on the overhead projector to show the different ways they made the bulb light.

LESSON 3

Figure 3-1

D-cell and Bulb

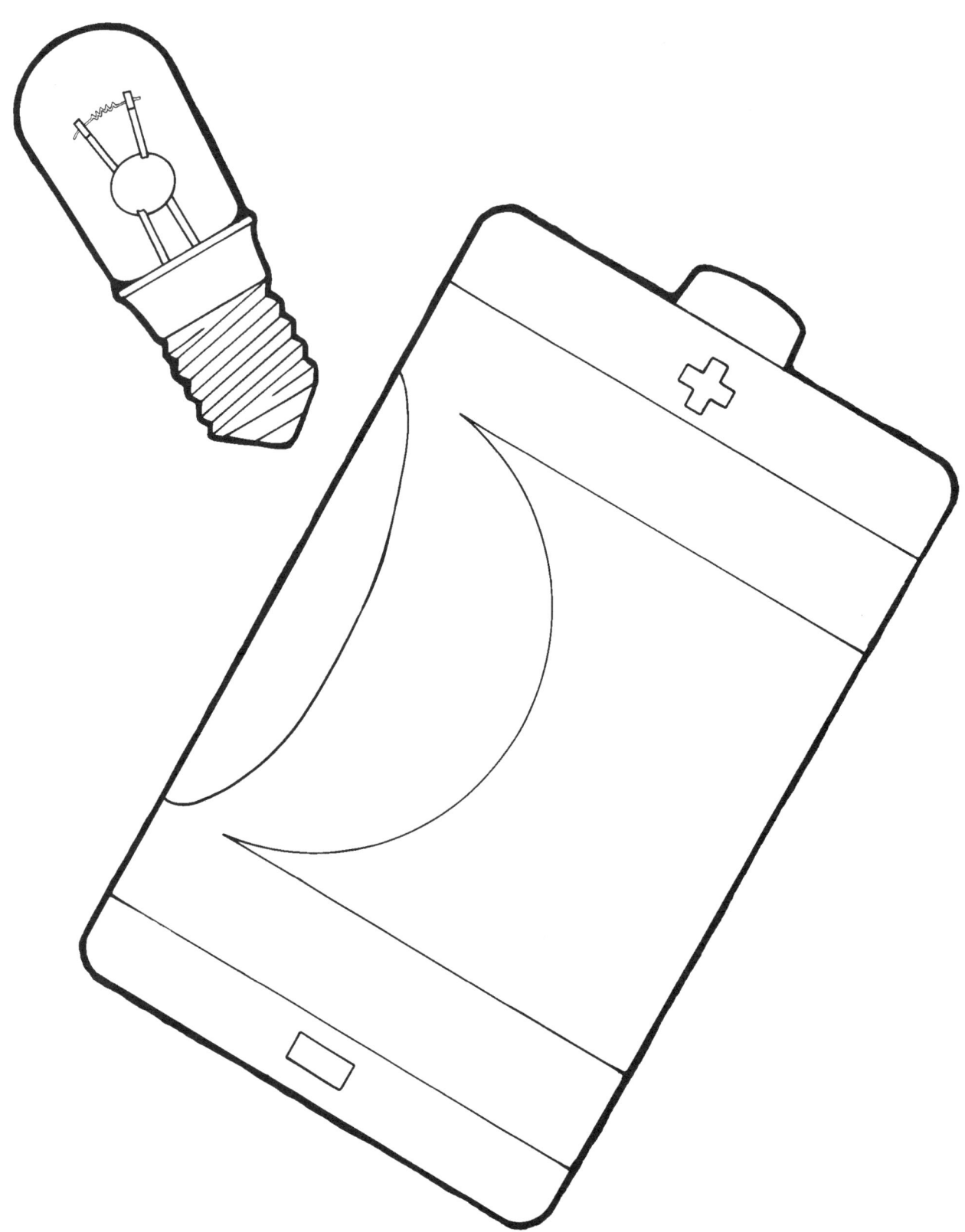

Procedure

1. Hand out **Activity Sheet 1**. Review the directions. Make sure students understand what they are supposed to do: to predict whether the bulb will light by writing "on" or "off" under each drawing.

    Assure them that they will not be graded on their "guesses." Instead, they are to use the predictions as a way to learn. You might emphasize that this is the way scientists work.

2. Give students time to work on the sheet. When everyone has finished, have students get their boxes containing the necessary materials. Have students check their predictions by connecting the wire and the D-cell in the configurations shown on the prediction sheet. Does the bulb light?

3. If students have mastered the circuits shown on **Activity Sheet 1**, ask them to come up with additional ones that work. Have them draw pictures in their notebooks of any new circuits. Urge them to make the drawings large enough to show the details of the wire, the D-cell, and especially the bulb.

4. Have the students put their materials back into their storage boxes and clean up, checking again to make sure they have only one of each item.

5. Have students go to the area where you displayed the large pictures and draw the wire to illustrate how they got the bulb to light.

Final Activities

Using either the large pictures or the transparency focus on the specific places on the bulb where the wire touches in order for the bulb to light. You might ask "Can anyone show us exactly where the wire needs to touch the bulb to make it light?" And then later ask, "Do any of your drawings show a different way to touch the wire to the bulb?"

Extensions

Suggest that students keep a running list of things they would like to learn about electricity. Also, have students record the new things they are learning as well as the new words they are using.

Evaluation

By the end of this lesson, you will have the following available to help you assess what students understand about electricity:

1. The brainstorming list of information about what your students know about electricity and their list of questions about what they wanted to learn about electricity.

2. Student notebooks containing a list of what they know about electricity, a drawing of a light bulb, and sketches of the various ways they could and could not get the bulb to light.

3. The student work on **Activity Sheet 1**.

4. In addition, your observations of students as they tried to light the bulb will give you information about their understanding of electric circuits, their ability to tackle new problems, and their learning styles.

Items 2, 3, and 4 give you information about individual students. Item 1 gives you information about the class as a whole.

A Closer Look at Circuits / **21**

LESSON 3

LESSON 3

Teacher's Answer Sheet
Activity Sheet 1

NAME: _____

DATE: _____

Will the bulb light or not? Below each picture, make your prediction by writing either "On" or "Off."

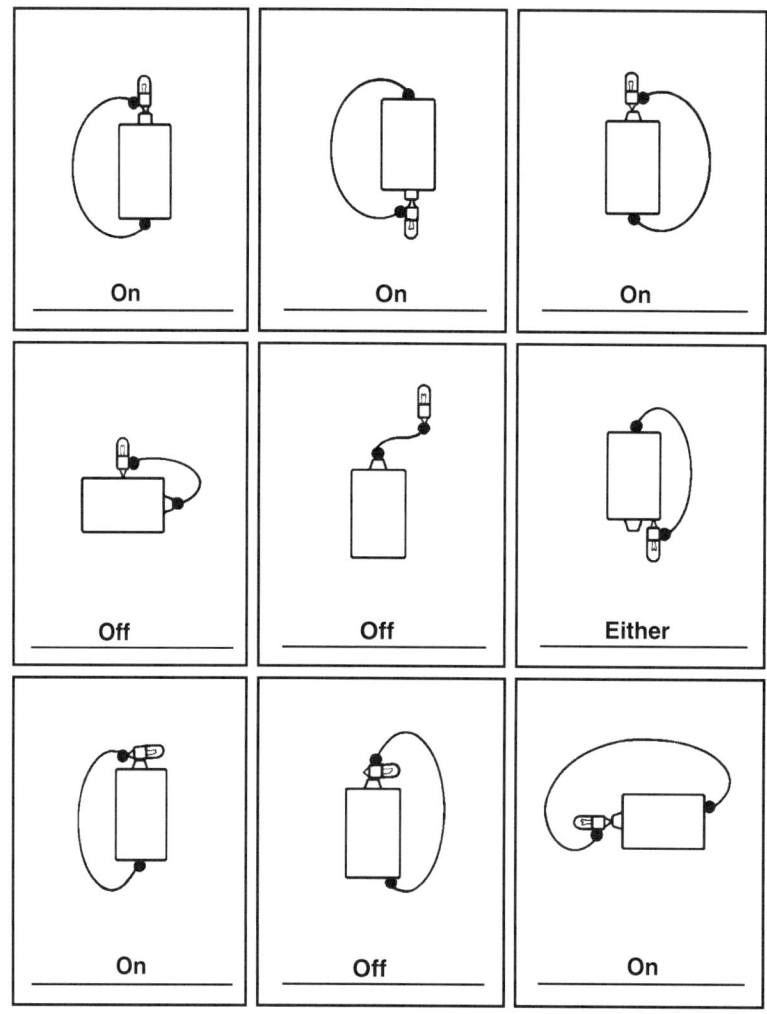

LESSON 3

Activity Sheet 1

NAME: _____

DATE: _____

Will the bulb light or not? Below each picture, make your prediction by writing either "On" or "Off."

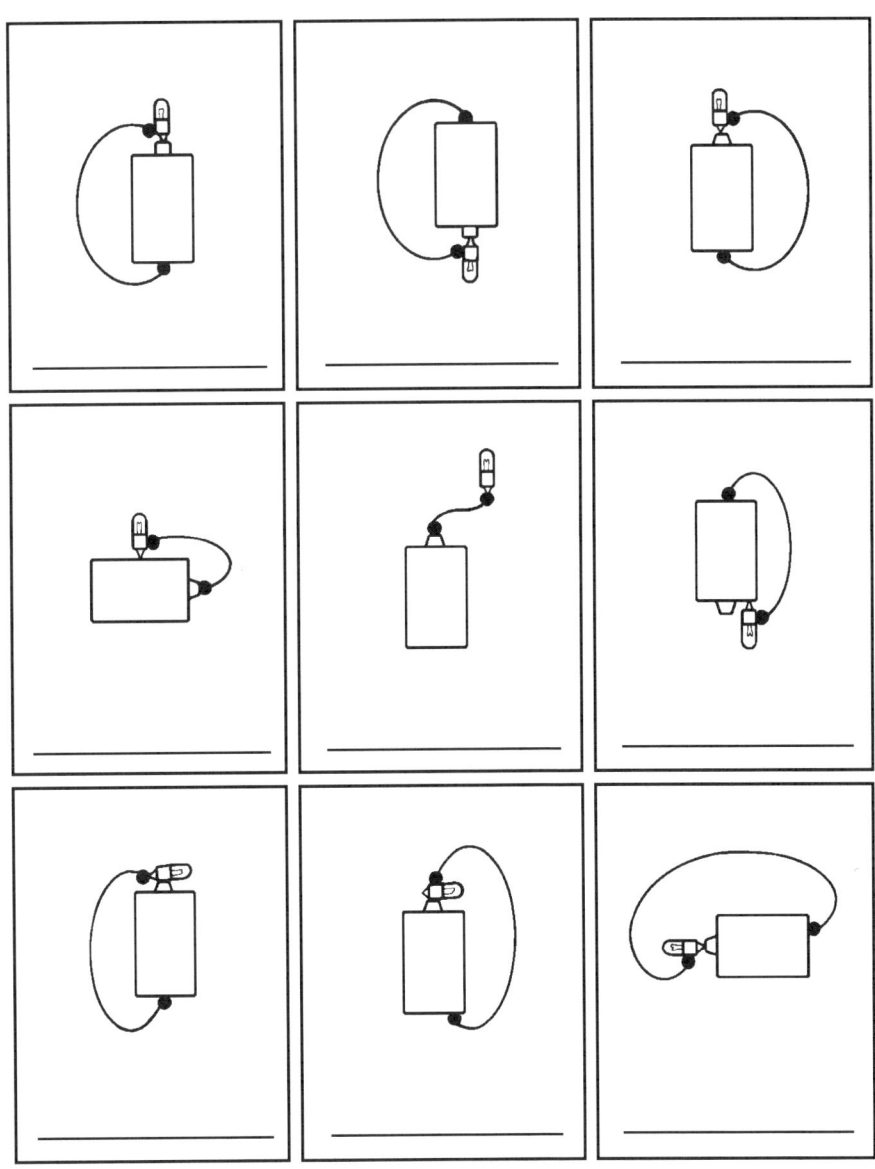

| LESSON 4 | # What Is Inside a Light Bulb? |

Overview

Light bulbs only produce light when they are part of a complete circuit. In this lesson, students will have an opportunity to see how to construct a circuit to light a typical household bulb. Under the direction of the teacher, the class will complete the circuit using a bulb and twenty batteries as a power source. The class will start out using five batteries and then increase the number to twenty; this will illustrate how adding batteries to the circuit will make the bulb glow more brightly. By using the large bulb, students will discover how a common household device works and will be able to see the parts of a bulb clearly.

Objectives

- Students further their understanding of circuits by constructing a curcuit to light a household bulb.
- Students learn to identify the parts of a bulb and to trace the path of electricity through a bulb.

Background

Figure 4-1 shows the parts of a 120-volt clear household bulb, which is fundamentally the same as the students' smaller bulbs. The only significant difference in the bulbs is that the filament (the part that lights up) of the large bulb is longer and is capable of producing a brighter light. The large bulb also requires 120 volts to light. By using twenty batteries in series, at 1.5 volts each, you provide about 30 volts. This will cause the filament to glow dimly.

Notice that a wire from one side of the filament passes through the glass and connects to the threaded metal base of the bulb. A wire from the other side of the filament passes through the glass and connects to the tip at the bottom of the base. The tip and the threaded base are separated by an insulating piece of ceramic. All bulbs share these features.

LESSON 4

Figure 4-1

Household bulb

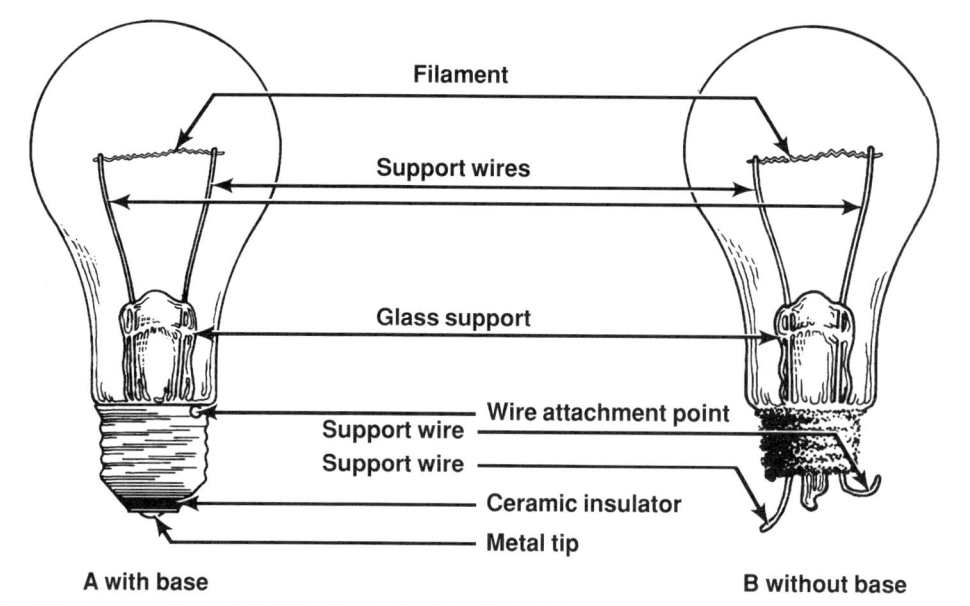

Materials

For each student
- 1 student notebook

For the class
- 2 yardsticks or meter sticks to help hold the batteries
  (The chalk tray on the chalkboard can be used instead.)
- 1 household bulb
- 1 household bulb with metal base removed
- 1 miniature bulb, with base removed

Note: If bulbs with bases removed are hard to get, you can proceed without them by simply lighting the standard bulb in the way described below, or you can remove the bases from the two bulbs yourself, before class. See **Appendix D**.

- 2 pieces of hook-up wire, each about 5 feet long
- 1 wire stripper
- 1 wire cutter
- 1 pair of needle-nose pliers
- 1 D-cell battery for each student

Preparation

Strip at least ½ inch of insulation off each end of the two 5-foot pieces of hook-up wire.

Procedure

1. Show the students a standard electric light bulb. Ask them how they think it can be lit using D-cell batteries and wires. Ask them to write their ideas in their notebooks.

2. After discussing how many batteries are needed to light the bulb, prepare the bulbs for a demonstration. Tell the class that you will need their help and cooperation.

26 / What Is Inside a Light Bulb?

2. After discussing how many batteries are needed to light the bulb, prepare the bulbs for a demonstration. Tell the class that you will need their help and cooperation.

3. Start with either the lowest student guess or with five batteries in the groove defined by the two sticks (Figure 4-2) or by the chalk tray. Make sure all the batteries point in the same direction. Next, assign three or four students to hold the batteries snugly together, as illustrated in Figure 4-2. Ask another two students to hold a wire in contact with the battery terminal at each end. Finally, have two more students hold the ends of the wire that touch the bulb.

Figure 4-2

Lighting a household bulb

4. While you hold the bulb, have two students place the ends of the wires against the bulb base so the bulb will light. Make sure the filament is visible to the class. If you darken the room, the filament will show more clearly when it does start glowing.

5. Add one battery at a time. When there are sufficient batteries in the groove (approximately fifteen), the filament will glow dim and red. With twenty batteries, the filament will be brighter but will still be quite dim compared to a lamp. However, the process will illustrate that the more batteries you add, the brighter the bulb becomes.

6. Give several students a turn at holding the wires and touching the bulb to make it light. Discuss with them where the bulb has to be touched to make it light.

7. If you are doing the second part of the demonstration using the large bulb with its base removed, now is the time to bring it out. Tell the students that the base has been removed to show what the inside of the bulb looks like (Figure 4-1, B). Show them the parts of the base that were cut away.

8. Using the same technique as above, light the bulb that has had the base removed by touching the lead wires from the batteries to the two wires that emerge from the glass on the bulb (see Figure 4-3).

Figure 4-3

Lighting the bulb

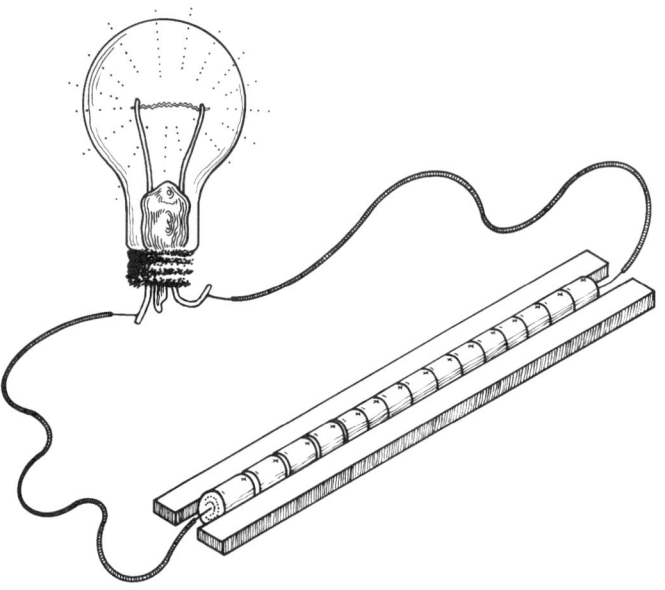

9. Point out the detail of the wire arrangement in the bulb. Notice the two straight wires that are rigid and pass through the glass up into the center of the bulb. The filament is stretched between the ends of these two wires.

10. Bring out the small bulb that has had the base removed. Compare the large bulb and the small bulb. The students will see that the design of the small bulb is similar to the large bulb.

11. Using two batteries, light the small bulb with the base removed in exactly the same way the large bulb was lighted: by using the long wires. Have students look for the filament. Can they see it?

Final Activities

Discuss how wires are connected to make the bulb light. Then ask students to draw the bulb and to indicate on their drawings where key wires are located.

Extension

Books on Thomas Edison and on the invention of the light bulb can be introduced at this point. Several are mentioned in **Appendix G**, and your library may have others.

28 / What Is Inside a Light Bulb?

| LESSON 5 | **Building a Circuit** |

Overview Now students have had enough experience working with circuits that they can think about building their own circuits. In this lesson, they will learn how to use some new devices to wire more permanent and complex circuits.

Objectives
- Students learn how to use a battery holder, a light bulb socket, and their attached Fahnestock clip—devices that will help them build circuits.
- Students gain more experience working with circuits.

Background This lesson marks the beginning of a series of lessons during which students will be building circuits and working with them. An electric circuit is a continuous path for electricity to travel in going from one end of a battery, through wires (including the wires in the bulb), back to the other end of the battery.

By convention, scientists and engineers speak of the electric current in a circuit as though it flowed from the positive end of the battery through the wires and back to the negative end of the battery. This convention arose because of Benjamin Franklin's theory that electric current was carried by positive charges. As it turned out, Franklin's theory has been proven incorrect; today we know that either positive or negative charges can move through an electric current. In metals, the negatively charged electrons are the ones that move.

When explaining the flow of electricity to students, focus on conventional current flow (from positive to negative). But, if students already know about electron flow and bring it up during the discussion, it should not be avoided.

Materials *For each student*
  1 storage box containing:
      1 D-cell battery
      1 bulb
      1 6-inch piece of wire

LESSON 5

    1   battery holder
    1   bulb socket
    1   6-inch piece of wire
    1   **Activity Sheet 2**
    1   student notebook

For the teacher
    1   pair of needle-nose pliers
    1   screwdriver

Preparation

1. Check the battery holders (see Figure 5-1) to see that they work. On some brands, the surfaces designed to make contact with the battery aren't positioned correctly. And sometimes the Fahnestock clips (located at the ends of each battery holder) may need adjustment. Use the needle-nose pliers to make adjustments.

Figure 5-1

Battery holder

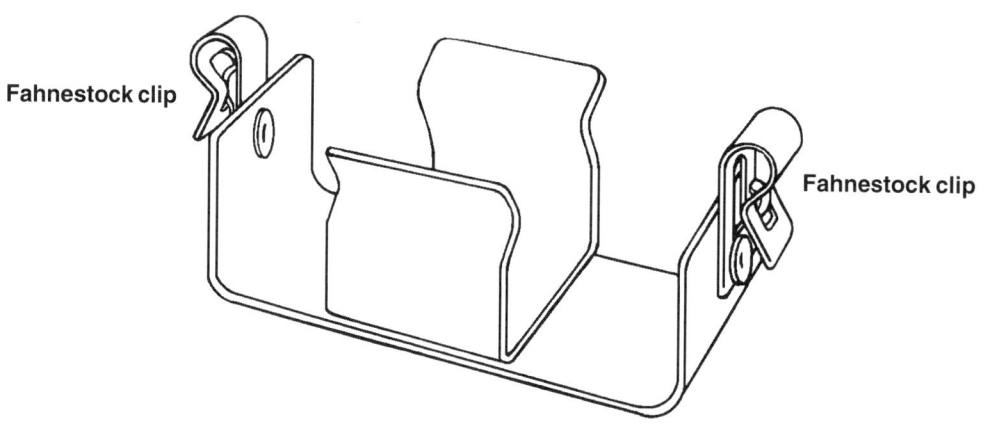

2. Prepare one more wire for each student by cutting and stripping the ends off additional 6-inch pieces of wire. If it is convenient, have a student helper strip the wire.

3. Use the large drawing of the Fahnestock clip in Figure 5-2 to make a transparency, or draw a picture of the clip on the chalkboard. Students can refer to the diagram if they are having trouble getting their clip to work.

4. Be prepared to spend more than one period to complete the lesson, depending on the speed with which students learn how to use Fahnestock clips, which are shown in Figure 5-3. Students frequently fail to push the wire entirely through the opening, so that when the spring is released the wire is not held in place securely. This problem often arises when the wire itself is not straight at the end.

Procedure

1. Distribute a copy of **Activity Sheet 2** to each student. Tell the students you would like them to predict which bulbs will light by writing "On" or "Off" on the line beneath each one. Have the students consult their notebook to make the predictions and then put the prediction sheet in their notebook.

30 / Building a Circuit

Figure 5-2

Fahnestock clip

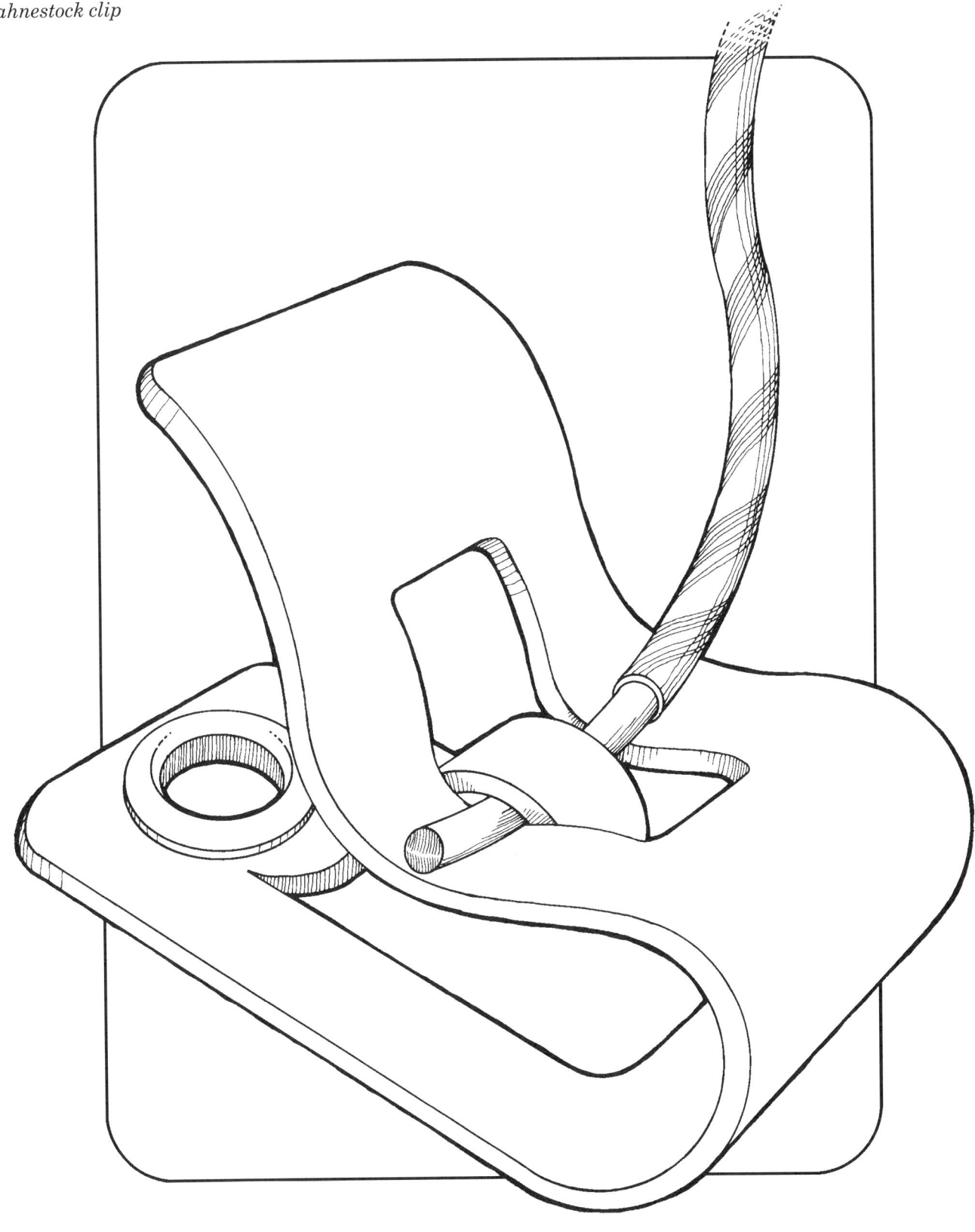

Building a Circuit / **31**

2. Next, give each student a battery holder and the second piece of wire. Point out the ends where the Fahnestock clips are located. Use a transparency of the large picture of the Fahnestock clip (Figure 5-2) to explain how to attach a wire to the Fahnestock clip at the end of the battery holder.

3. Make sure everyone can connect a wire to the Fahnestock clip. Have students work in pairs and encourage them to help one another use it correctly. This is worth the effort, because making circuits will be much easier if students are comfortable using the Fahnestock clips. Figure 5-3 illustrates how to insert the wire into the clip.

Figure 5-3

Using a Fahnestock Clip

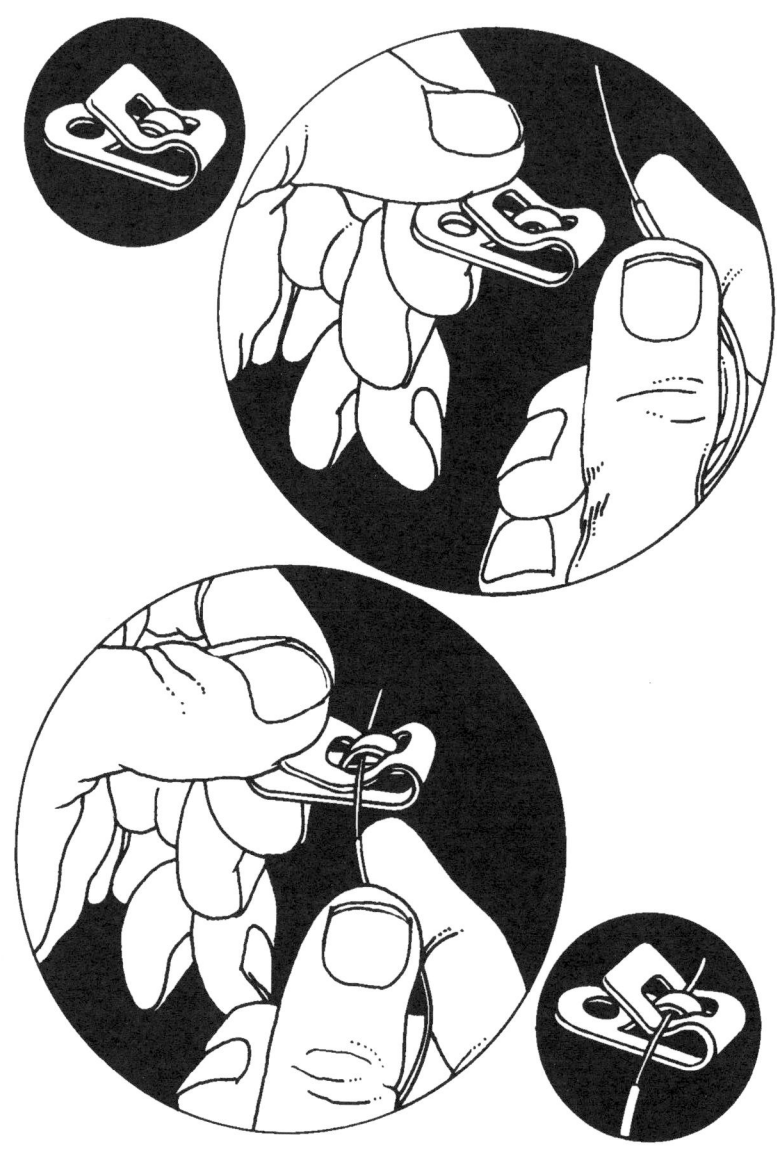

4. After students have mastered using the Fahnestock clips, have them place the batteries in their holders and their wires on the Fahnestock clips. When everyone has a battery in place and a wire attached to each end, divide the students in pairs. Have one student hold the bulb taken from the box of materials. Have the other student hold a piece of wire in each hand and use the wire to touch the bulb at either end until it lights. As

32 / Building a Circuit

students try to light the bulb by touching the tip of the wire to the correct part of the base of the bulb, they will reinforce what they learned in Lesson 4.

5. Next, give each student a bulb socket (Figure 5-4) and challenge them to connect the entire system—the battery in the holder, the bulb in the socket, and the wires—so that the light stays on (see Figure 5-5). Encourage students to work at their own pace until they complete the task.

Figure 5-4

Bulb socket

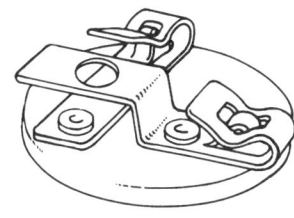

Figure 5-5

Complete circuit

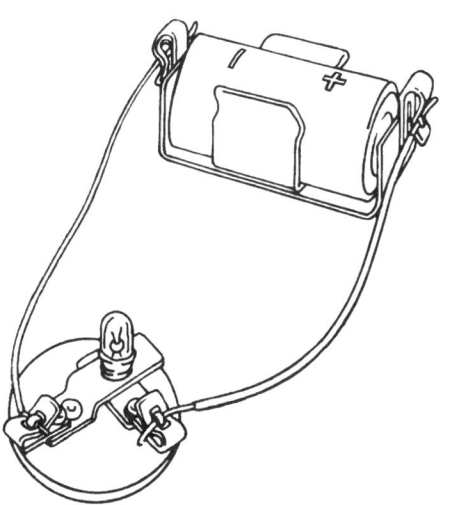

6. Make sure each student can accomplish the task. Have students work together so that everyone gets the system to work. If the students are grouped in teams of four, tell the teams you want them to help each other build the circuit.

7. Tell students to put their materials back in their storage boxes. Tell them to make sure the circuit is not connected.

Final Activities

End the class with a discussion. Ask the students to describe the way the battery, bulb, and wires must be connected to have the bulb light. As students describe the arrangements, draw several representative circuits on the board. With a few working circuits drawn on the board, ask the students to look at them and tell you what they have in common. Ask: "What are the crucial connections and elements necessary to make the bulb light?"

End the discussion by drawing out from students the idea of a circuit: the path of the electric current as it goes from one end of the battery, through the wire, through the bulb, through the next wire, and back to the other end of the battery.

LESSON 5

Extensions In a learning center or on a table in the room where students can work during free time, have available several batteries, bulbs, holders, sockets, and wires. When students create new circuits, ask them to draw these circuits in their notebooks.

Evaluation
1. Review the student responses to **Activity Sheet 2**. Students who have not been able to accurately predict which bulbs will light may need some small group instruction.
2. Observe how students progress as they attempt to assemble the circuit. Students who are struggling may benefit from working in a small group, where you can focus directly on the things they do not understand.
3. The drawings that students made will give you information about how clearly the students see the parts of the circuit, and whether they can draw the circuit as continuous, starting from one end of the battery, going through the wires and the bulb, and returning to the other end of the battery.

LESSON 5

Teacher's Answer Sheet
Activity Sheet 2

NAME: _____

DATE: _____

Will the bulb light? Below each picture, make your prediction by writing either "On" or "Off."

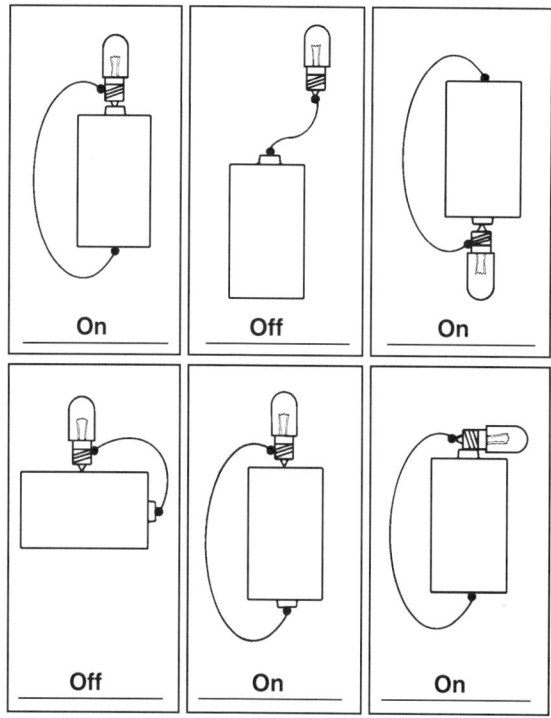

34 / Building a Circuit

LESSON 5

Activity Sheet 2

NAME: _____

DATE: _____

Will the bulb light? Below each picture, make your prediction by writing either "On" or "Off."

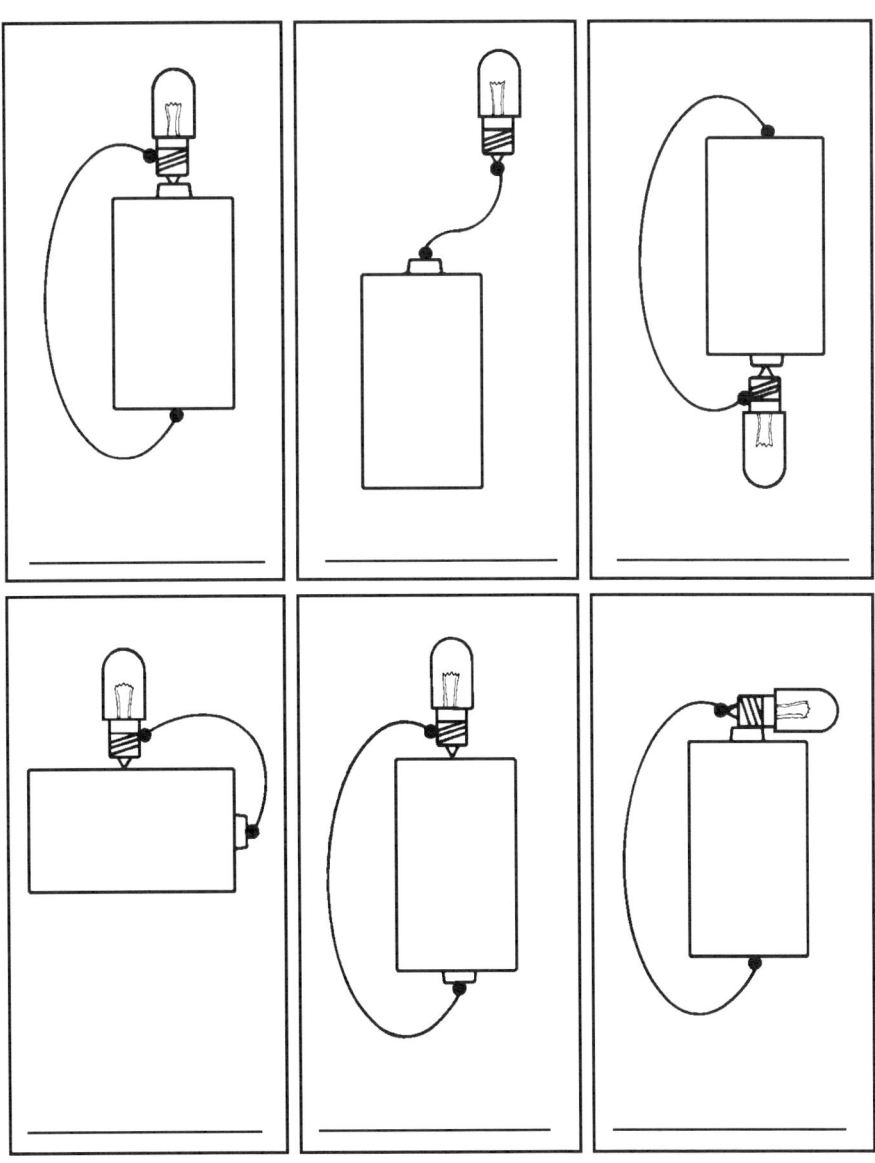

| LESSON 6 | # What's Wrong with the Circuit? |

Overview In all facets of life people face problems that need to be solved. Sometimes the problem may be how to construct something new. At other times, it may be finding out how to fix something that isn't working. But whatever the problem, there are certain strategies that can make the job of problem-solving easier. In this lesson, a step-by-step problem-solving technique is developed that students can use to test electric circuits.

Objectives
- Students build a circuit tester.
- Students think about different ways to use their circuit testers.
- Students learn a troubleshooting technique to check their circuits.

Background This lesson focuses on an important problem-solving skill. With teacher guidance, students will help develop an effective troubleshooting technique. This technique will give students practice in isolating the elements of a circuit and figuring out, by a process of elimination, which part is malfunctioning. Mastering this technique will help students resolve problems without constantly asking the teacher for assistance.

Materials *For each student*
  1 storage box containing:
      1 D-cell battery
      1 battery holder
      1 bulb
      1 bulb socket
      2 6-inch pieces of wire
  1 6-inch piece of wire
  1 **Activity Sheet 3**
  1 student notebook

LESSON 6

For the teacher
- 6 6-inch pieces of wire
- 2 batteries in holders
- 1 bulb in a socket
- 1 faulty bulb in a socket

Preparation

1. Make the two versions of the circuit tester illustrated in Figure 6-1. One of them should contain the faulty bulb. Test both by touching the two end wires together to see if the bulb lights in the functioning circuit.

Figure 6-1

Circuit tester

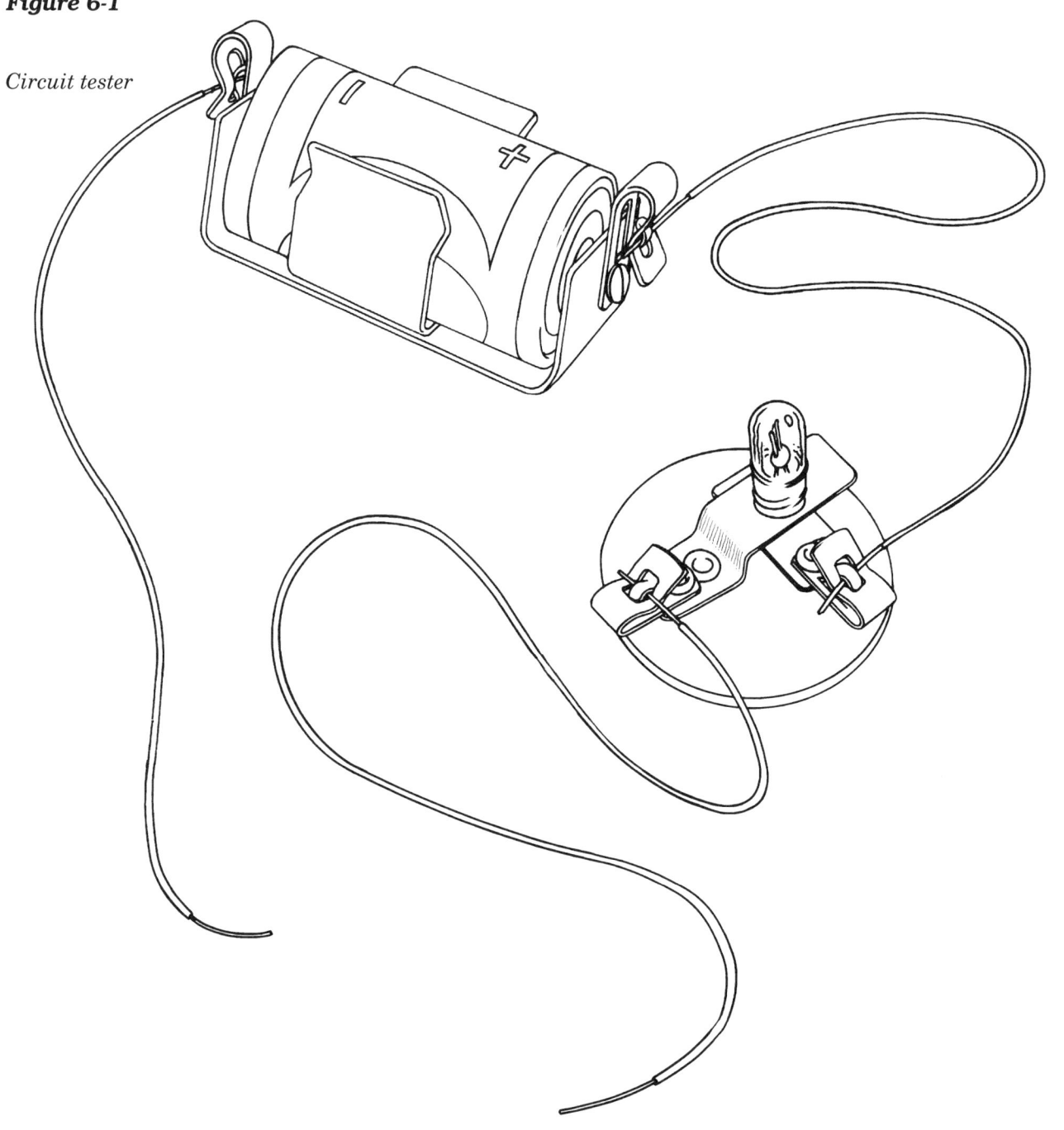

38 / What's Wrong with the Circuit?

LESSON 6

2. Draw the circuit tester on the chalkboard or on a large piece of newsprint.

   You will use the circuit tester containing the faulty bulb to demonstrate to the class how to systematically go about finding the faulty part.

3. Prepare another wire for the class.

4. Photocopy a class set of **Activity Sheet 3**.

Procedure

1. Begin the class by giving students **Activity Sheet 3** and asking them to complete it. Collect the sheets before proceeding.

2. Next, tell the students you want to talk about solving problems. For example, discuss how to fix a gadget or a circuit that isn't working. Encourage students to reflect on the tactics they might use.

3. Demonstrate the two devices you have prepared. Show the students the one with no faults in it. Touch the two wire ends together so the students can see that the bulb lights. Then, using the device with a faulty bulb or a faulty battery in it, touch the two wire ends together so that the students can see that the bulb does not light.

4. Ask the students how they would go about figuring out how to make the bulb in the second circuit light. After allowing some time for students to think, ask different ones to share the way they would proceed to try to figure out what is wrong with this circuit.

5. Using the student's responses, make a list on the board of the steps for finding out what is wrong. Introduce the term **troubleshooting**—a way to find out what is wrong. The list might include the following steps. This list has *not* been included in the Student Activity Book. Offer these suggestions to students, if necessary. Use your own student-generated list if possible.

   - First, look at the wire connections to make sure they are secured tightly. Do this by wiggling them gently to check that they are not loose. Test the circuit (by touching the end wires together) again to see if the bulb lights now.

   - Look at the battery and make sure it is in its holder tightly and that the ends of the battery are touching the battery holder. Test it again (by touching the end wires) to see if the bulb lights.

   - Next, look at the bulb and make sure it is screwed securely into the socket. (But not too much. Students often keep turning the bulb more than is necessary to make contact.) Test the circuit again to see if the bulb will light.

   - If it still doesn't work, check the battery and the bulb separately by first taking out the bulb and lighting it using one wire (as in Lesson 2) and a good battery.

   - If the bulb is good, the next step is to test the battery. Take the battery from the holder and test it again by using the one-wire technique (as mentioned next to the first bulleted item) with a bulb that you know is good.

   - At this point, if you have established that the bulb *and* the battery are good, put the system back together. If it still doesn't work, you know there must be a loose connection somewhere in the system.

What's Wrong with the Circuit? / **39**

LESSON 6

In summary:

- Are the wires placed properly in the Fahnestock clips?
- Are the wires firmly attached to the battery holder?
- Are the wires firmly connected to the bulb socket?
- Is the bulb burned out?
- Is the battery worn out?

6. Give each student one more 6-inch wire stripped at each end. Ask the students to assemble their own testing circuit, using the materials they have in their boxes. Draw the device on the board and leave your circuit tester to give them a model to follow, as well.

7. Have students put their materials in their storage boxes, store them in the designated place in the class, and clean up.

Final Activities

Review the troubleshooting process you have established. Define the term *troubleshooting* as a technique for solving problems. Remind students that they can use this technique whenever they think a piece of their equipment is not working properly. Tell them what you want them to do when they find that either a battery or bulb is faulty. (For example, they might just bring it to you to get a replacement.)

Extensions

1. In the learning center or on a table, place three or four circuits, each of which has a different fault in it: a loose connection, a burned-out bulb, or a drained battery, for example. Number each circuit. Challenge students to figure out what is wrong with each one.

   Remind students to record their findings in their notebooks. Tell students not to reveal their solutions to others so all students can have the fun of solving the puzzles.

2. Ask students to interview one of their parents about problem solving. They might ask how the parent solved a problem around the house or with the car. Suggest that students write a brief account of the strategy that was used in their notebook.

Evaluation

1. Review student responses to **Activity Sheet 3**. Invite students who need further work to use their equipment to confirm their predictions.

2. Observe students when they create their own circuit testers. This will tell you whether students understand how to assemble the materials they have at this point.

   You will notice some students who have not mastered working with the Fahnestock clips or with some other part of the equipment. You might organize some "expert" students (possibly those students who are usually not successful in school) to serve as helpers for students who haven't gotten it yet.

3. While the students are working in groups to make their circuit testers, observe the way the groups are functioning. Are they working cooperatively? Are the faster members of the group being patient and appropriately helpful to the ones who are having difficulties? Group

40 / What's Wrong with the Circuit?

techniques don't come naturally. Students need to be taught and constantly coached to learn how to work together in ways that show respect and consideration.

LESSON 6

Teacher's Answer Sheet
Activity Sheet 3

NAME: _____

DATE: _____

First, draw one wire on the top two pictures so that the bulb will light. Then complete the partially drawn wires in the bottom two pictures to make a complete circuit, which will light the bulb.

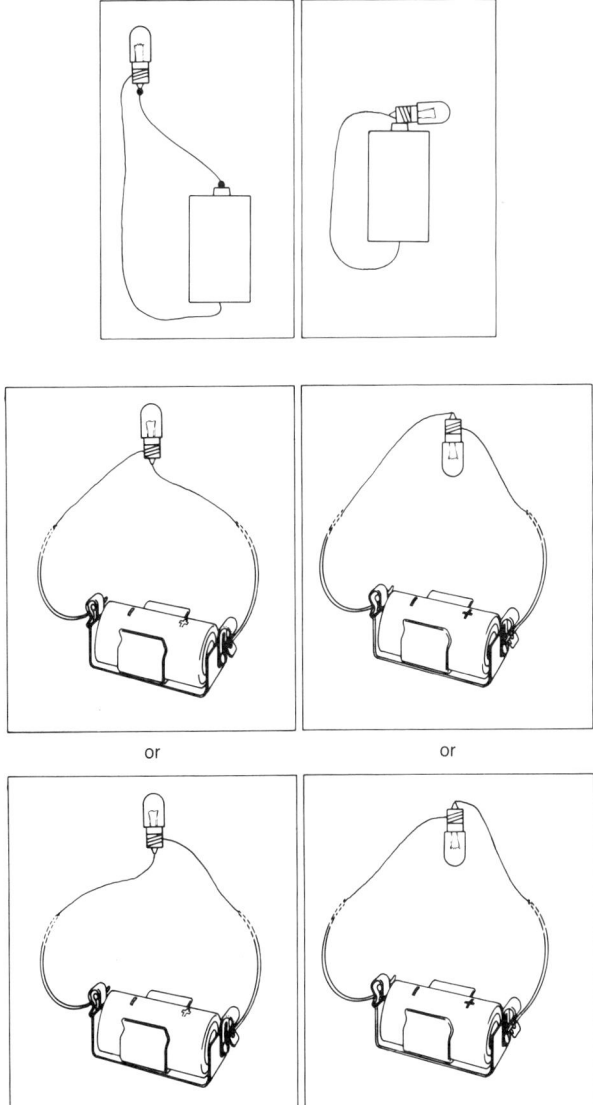

or   or

What's Wrong with the Circuit? / **41**

LESSON 6

Activity Sheet 3

NAME: _____

DATE: _____

First, draw one more wire on the top two pictures so that the bulb will light. Then complete the partially drawn wires in the bottom two pictures to make a complete circuit, which will light the bulb.

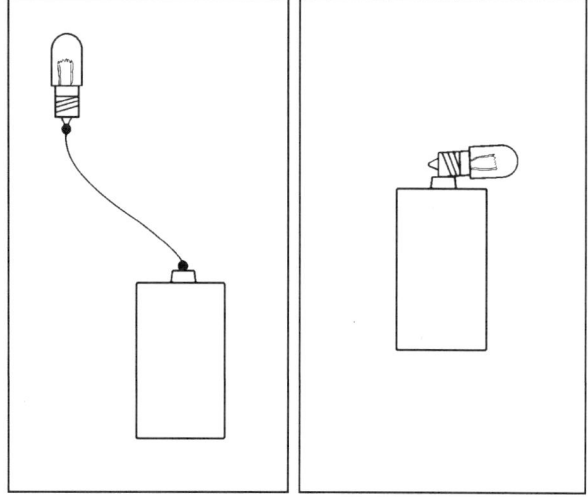

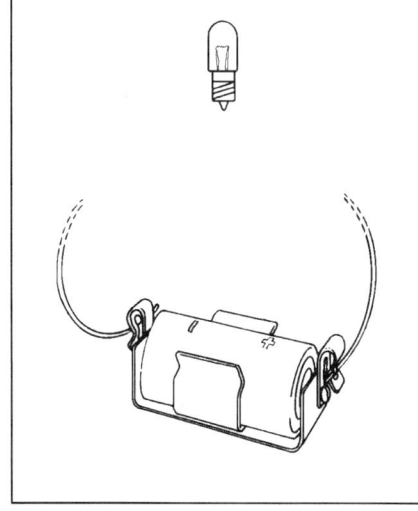
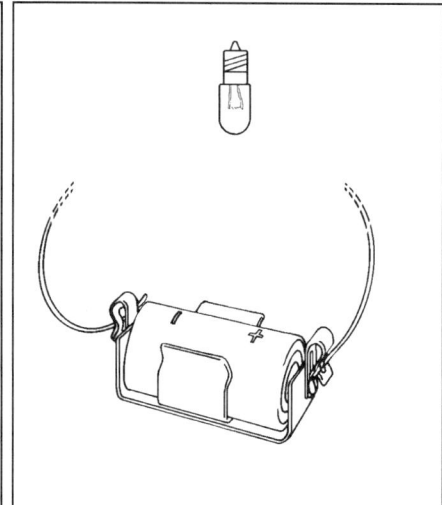

LESSON 7

Conductors and Insulators

Overview

At this point in the unit, students have gained some understanding of circuits and how they work. For the next two lessons, they will be exploring the parts of a circuit. In this lesson, they will discover that **conductors** are materials that allow electricity to pass through them. They will also learn that **insulators** are materials through which electricity does not flow—at least not in detectable amounts.

Objectives

- Students develop an understanding of the behavior of electrical conductors and insulators.
- Students learn how to use a circuit tester to identify conductors and insulators.

Background

Conductors and insulators are important components in electric circuits. Conductors are materials through which electricity can travel. Insulators, on the other hand, are materials through which electricity cannot travel.

To determine if a material is a conductor or an insulator, students will be asked to place the material between the two wires of the circuit tester. Figure 7-1 shows how to use a circuit tester to see if an object can be part of a circuit. If the bulb lights, then students will know that the material conducts electricity.

Figure 7-1

Using a Circuit tester

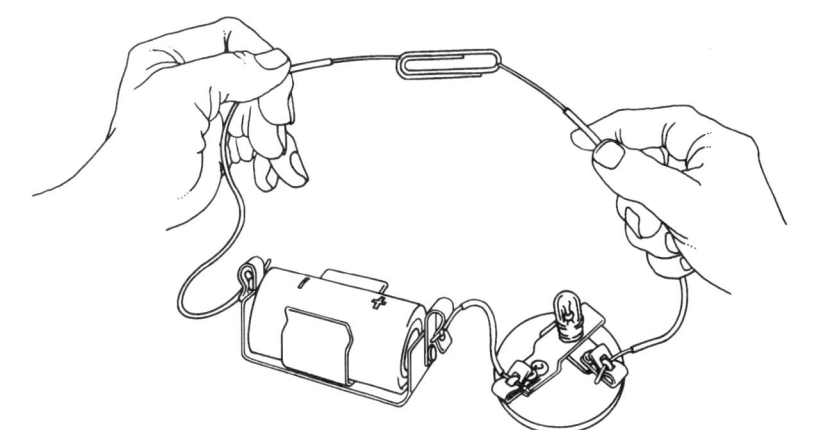

Conductors and Insulators / **43**

LESSON 7

As you might suspect, the classification of materials as *conductors* and *insulators* can get more complex. For example, there are materials called **semiconductors** that sometimes act as conductors and at other times act as insulators. Later in the unit, students will work with a semiconductor diode like those used in computer chips. But in this lesson, it will be sufficient to classify materials as conductors or insulators.

The package of assorted materials specified in the Materials List (see **Unit Overview**) should give consistent results. But don't be surprised if some of the materials that you know are conductors show up on a student's list as nonconductors. The coatings on some metals (like bottle tops) may prevent good electrical contact, causing students to classify some metal items as nonconductors. To rectify this problem, the students can use a nail or sandpaper to scrape off the coating. Also, different parts of an item might be conductors or insulators; for example, the metal eraser holder on the pencil is a conductor, while the wood of the pencil is not.

Materials

For each student
- 1 circuit tester
- 1 student notebook

For every two students
- 1 package of the following assorted objects:
    - 1 golf tee
    - 1 1-inch piece of soda straw
    - 1 brass screw
    - 1 paper clip
    - 1 piece of aluminum screening (1 inch square)
    - 1 piece of plastic screening (1 inch square)
    - 1 1-inch piece of chalk
    - 1 wooden pencil stub (no eraser, lead exposed at both ends)
    - 1 brass paper fastener
    - 1 wire nail
    - 1 aluminum nail
    - 1 marble
    - 1 1-inch piece of pipe cleaner
    - 1 1-inch piece of bare copper wire
    - 1 1-inch piece of bare aluminum wire

For the teacher
- Chalkboard space (or large pieces of newsprint and a marker)
- 1 circuit tester (from Lesson 6)
- 1 sheet of fine sandpaper

Preparation

1. Sharpen the pencil stubs so that graphite is visible at both ends.
2. Assemble the materials above into small containers that can be handed out to each pair of students. These items are just suggestions. You may wish to add or substitute other items. You could have students collect small items to test at home and bring them to school in a small bag.

Procedure

1. Have students remove and check the circuit testers that they constructed in Lesson 6. Have students make sure that they work properly by touching the ends of the wires together to make the bulb light, as shown in Figure 7-2.

Figure 7-2

Testing the circuit tester

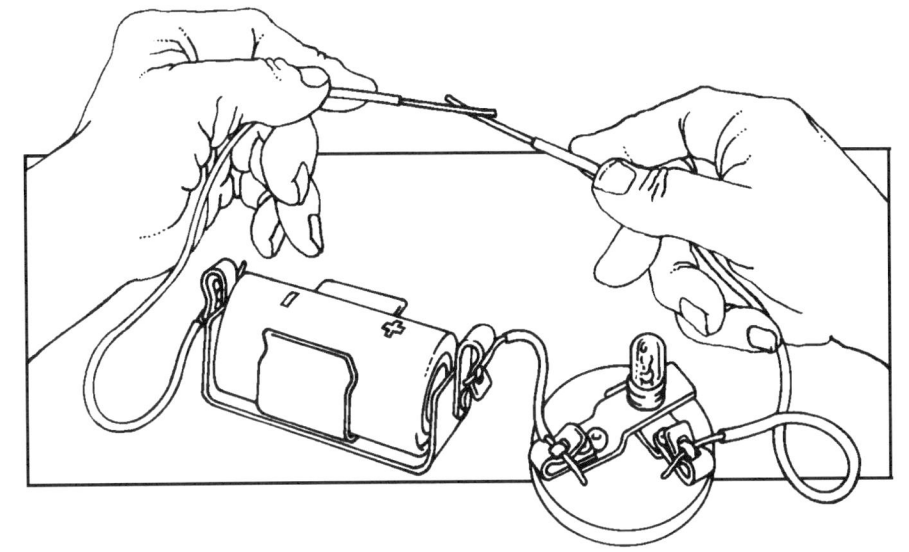

2. Ask the students how to use the circuit tester device to see if an object, such as a paper clip, can be part of the circuit. Some student will probably suggest something like the arrangement shown in Figure 7-3. Ask them: "What does it mean if the bulb lights when you touch the wires to the ends of the paper clip?" Help students see that if the bulb lights, it means the electricity had to travel through the paper clip. Use your circuit tester on something that does not conduct, such as a piece of chalk, and discuss why the bulb does not light.

Figure 7-3

Testing the paper clip

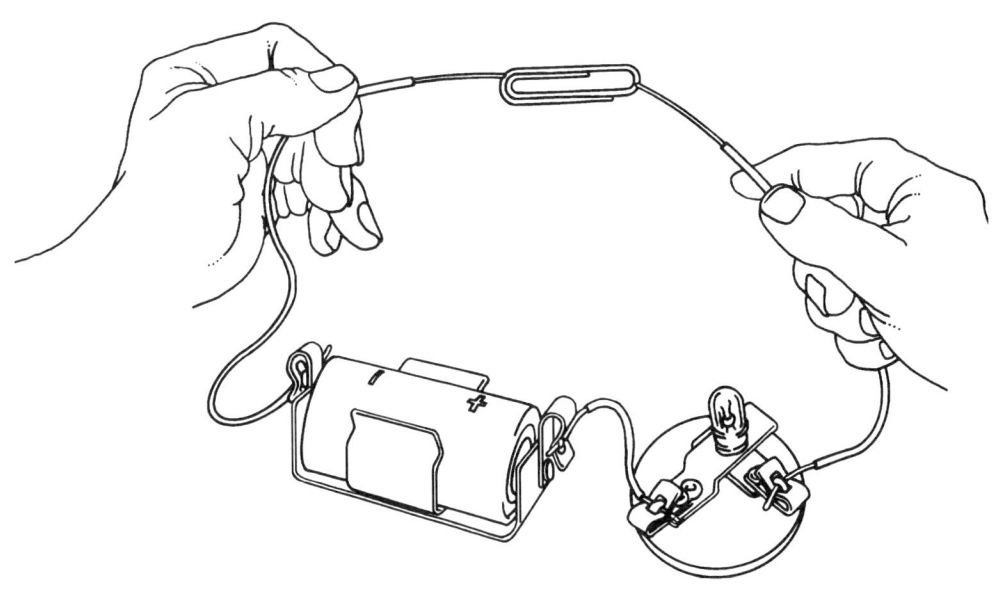

Conductors and Insulators / **45**

LESSON 7

3. Have each student take their set of assorted materials and create a list of these items in their notebooks.

4. Using the list, ask students to make a chart with three columns beside the list, and to predict whether they think the light will be on or off when the item is included in the circuit. An example is shown below.

Item	Predictions		Experiment Result
	Light On?	Light Off?	

5. Have each student use their circuit tester to test their set of assorted materials. Encourage the students to test materials they find in their desks as well. Emphasize the importance of recording the things that don't cause the bulb to light, as well as those that do.

6. About 5 minutes before the end of the period, have students return their items to their boxes and return the materials to the storage area.

Final Activities

1. At the top of two columns on the chalkboard, write: "Did Light" and "Did Not Light." Leave enough room to insert another heading above each of these titles. Ask students to tell you all the items they tested that caused the bulb to light. Record these in the appropriate column. When you get a "repeat," put a mark after the item to indicate that it has been mentioned before.

   Expect a few contradictory results. These provide an excellent opportunity to encourage students to resolve their differences by doing more experiments. If possible, a student might do an experiment to resolve the disagreement.

2. When the lists are complete, ask students to record them in their notebooks. Tell students that there is one word that describes materials that cause the bulb to light. That word is **conductor**. Write "Conductor" above "Did Light." Ask the students to define conductor. Using the students' words as much as possible, develop a class definition.

   Ask students to tell you the items they tested that did not light the bulb. Then explain to the class that there is a word that describes materials through which electricity cannot pass. That word is **insulator**. Write "Insulator" above "Did Not Light." Using students' language, develop a class definition of insulators.

3. Have students write the words "insulator" and "conductor" in their notebooks under their lists, and the class definitions, then have them draw a conductor and an insulator next to each definition.

Extensions

1. Encourage students to continue to investigate different materials and to keep lists in their notebooks of insulators and conductors.

2. Test the different parts of the bulb socket. Are the plastic base and the metal of the Fahnestock clips conductors or insulators? Then test the parts of the battery holder. Which parts are conductors? Which parts are insulators?

3. Tell students you want them to think about the ways electricity is used and transmitted to their homes. What parts of the system are insulators? conductors? For example, what about power lines, power poles, or the glass knob used to support the wires? Ask students to keep a list of their findings in their notebooks and to share their ideas with the class.

> **Safety Reminder**
>
> Tell students **never** to put their test wires into electrical wall sockets.

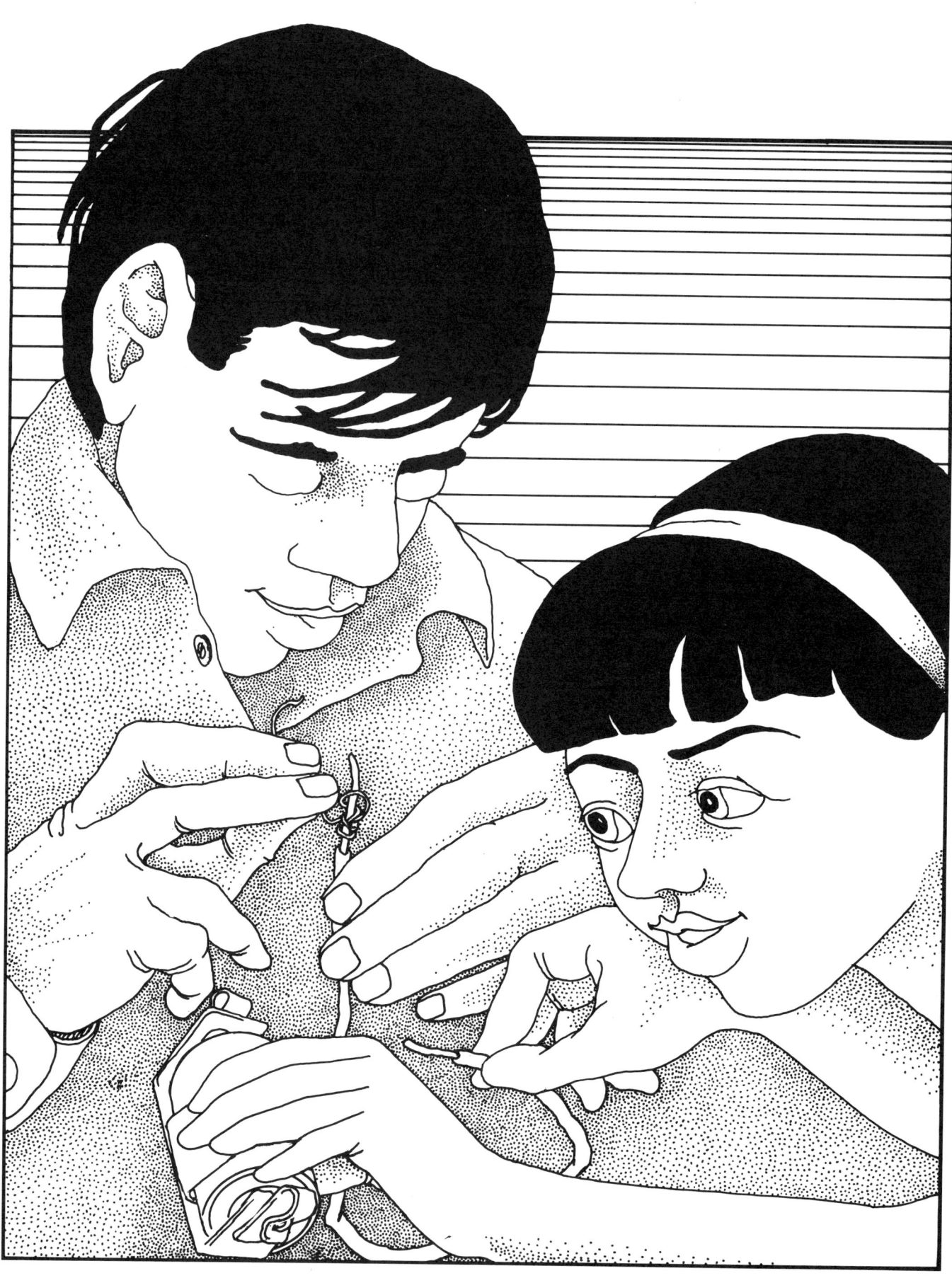

| LESSON 8 | # Making a Filament |

Overview

The filament, the piece of wire that gives off light in a light bulb, is an important element of the circuits students have been creating. In this lesson, they will explore the structure of a light bulb in more detail. They will make a device similar to a light bulb that includes a filament constructed from a piece of nichrome wire.

Objectives

- Students construct a device similar to a light bulb.
- Students learn that electricity can be used to generate heat and light.

Background

As you saw in Lesson 4, a household bulb consists of two rigid conducting wires that pass through a glass envelope. Inside the glass, a filament is stretched between the two conductor wires. When an electric current is passed through the filament, it gives off light and heat.

The filament in a bulb is made of an alloy of metals that is designed to last a long time. To prolong the life of the filament, the bulb is filled with an inert gas. This is needed because the oxygen in air would cause the filament to burn up quickly.

In this experiment, students use nichrome wire, made of an alloy of chromium and nickel, to make a filament. The nichrome glows and gives off considerable heat when a current is passed through it.

Materials

For each student
- 1 storage box containing:
  - 1 D-cell battery
  - 1 battery holder
  - 1 bulb
  - 1 bulb socket
  - 3 6-inch pieces of wire
- 1 student notebook

Making a Filament / **49**

LESSON 8

For every two students
- 1 ball of clay
- 1 4-inch piece of #32 nichrome wire

Preparation

1. Try the experiment yourself ahead of time to see how it works. Follow the instructions in the **Procedure** section. Stick the clay to a table so that the wires are held in a vertical position, as in Figure 8-1. Do not get clay on the bare ends of the wire because it will prevent good electrical contact.

Figure 8-1

Model of student-constructed light bulb

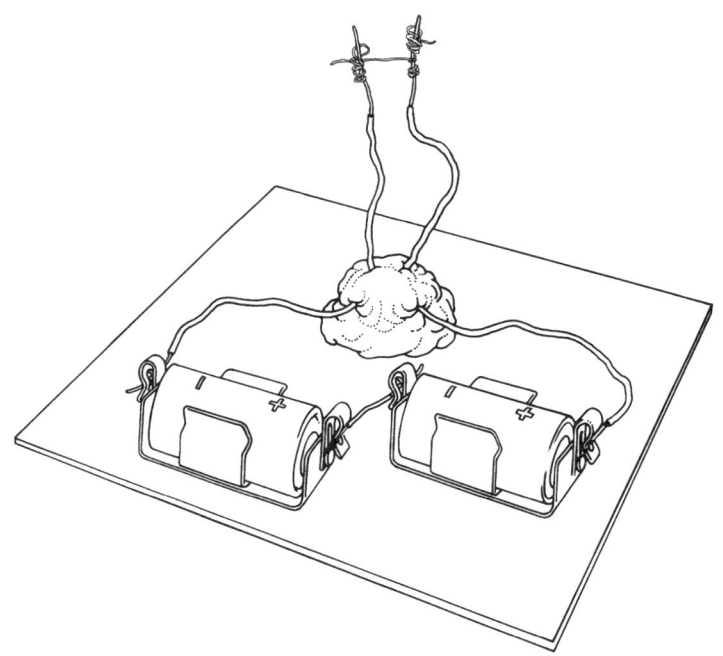

2. Read the safety reminder in the box below. Now wrap a 4-inch piece of #32 nichrome wire between the ends of the wires. You will need to pull the nichrome tight to get a good connection. There should be only about ½ inch of nichrome between the two ends of the hook-up wire.

Safety Reminders

The thin nichrome wire can cause small cuts on the fingers if it is pulled too hard.

The nichrome wire gets hot when it glows. Remind students not to touch it, even if it falls off, until it has cooled.

The hot nichrome wire can ignite flammable materials, such as paper.

---

3. To make the nichrome wire glow, connect two batteries, in series, across the other ends of the wires, as shown in Figure 8-1. You will use this device to demonstrate the procedure for the class.

4. For this lesson, the students will work in pairs. Decide how you want the students paired.

Procedure

1. Remind the students of the work they did in Lesson 4 when they lit the household bulb and looked at its parts. Ask them to look back in their notebooks to see the drawings they made of the bulb. Also, remind them that a complete circuit is needed to light the bulb. Review, too, the uses of conductors and insulators in constructing a circuit.

2. Tell the students: "Now we are going to make a device that is like a light bulb in some ways."

3. Demonstrate to the class how to make the device (Figure 8-1). The steps are listed below:

    - First, divide the lump of clay into two parts. Place one part on the desk. Take two of the wires from the student kit (6 inches long each, with insulation stripped off each end) and put them on the clay so they are about ½ inch apart, with the middle part of the wire pushed into the clay.

    - Place the second lump of clay over the wires and knead the clay so that it supports the wires. Pull the top ends of the wire upright. Wrap the nichrome wire to the upper ends of the bare hook-up wire. Attach the two batteries to the free ends of the wires.

    - Warn students that the nichrome wire will get hot, so they need to attach it to the bare hook-up wire in such a way that they will not need to touch it when it is heated. Repeat out loud to the students all the safety reminders listed in the **Preparation** section.

    Note: If the filament doesn't light, students should make sure that they have made good connections, the wires are clean, the batteries are lined up in the same direction, and the nichrome wire is short enough.

4. Distribute the materials and instruct students to make their own devices. The students will need to use both of their batteries and connect them in the way shown in Figure 8-1.

5. Have students draw the devices in their notebooks, showing the part of the nichrome wire that glowed.

6. About 10 minutes before the end of the period, begin cleanup. Have students let the nichrome cool for about a minute. Then have students pull their wires from the clay, return the wires to their boxes, and return the clay to a location you have designated.

Final Activities

Ask students what they think the glass part of a light bulb does. After some discussion, ask whether they think the glass is a conductor or an insulator. (The glass acts as an insulator.)

When everyone has expressed an opinion, ask: "How can we find out?" Some students might suggest testing the glass with a circuit tester. After testing the glass, ask students to summarize what they have learned.

Extensions

1. Make available to students some books about the invention of the light bulb and about the life of Thomas Edison. Two books are listed in **Appendix G**. These are fascinating stories, worth pursuing.

2. Have students write a story about a light bulb. Tell them to include all the parts they have identified.

Making a Filament / **51**

| LESSON 9 | **Hidden Circuits** |

Overview

This lesson gives students a chance to apply what they have learned about circuits to a new problem. Using an enclosed box that has hidden circuits wired inside, students locate the circuits by using their circuit testers. In carrying out this task, students will develop some useful problem-solving strategies.

Objectives

- Students use a circuit tester to locate hidden conductors.
- Students develop confidence in their own problem-solving skills.

Background

The hidden circuit boxes present an interesting problem for students to solve. They must use their circuit testers, and they will need to develop a systematic strategy to determine the hidden wiring pattern. In the process, they will make use of the knowledge of circuits gained in previous lessons.

Materials

For each student
    1 storage box containing:
        materials for circuit tester
    1 student notebook

For every two students
    1 hidden circuit box

Preparation
(3 hours minimum)

1. You will need to construct a hidden circuit box for every two students. The instructions for constructing the boxes are in **Appendix E**. The construction of the boxes will require a minimum of 3 hours, so you may want to organize volunteer or student helpers to do the job.

2. Be sure each contact point on the top of the box is numbered, as in Figure 9-1. It will be very useful to have these numbers written on the underside of the lid, as well.

LESSON 9

Figure 9-1

Hidden circuit box

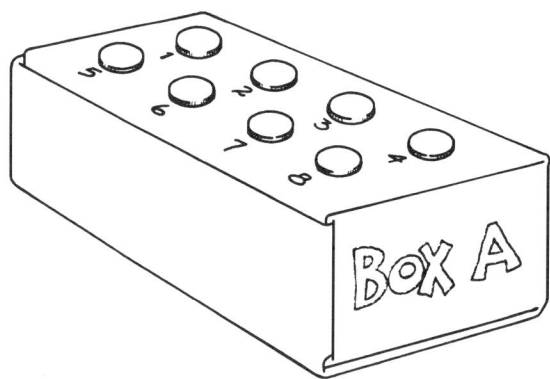

3. Check the condition of the hidden circuit boxes by opening the lid. Each box should contain two or three wires, and each wire should connect two of the contact points. Figure 9-2 shows the inside of a hidden circuit box.

Figure 9-2

Inside a hidden circuit box

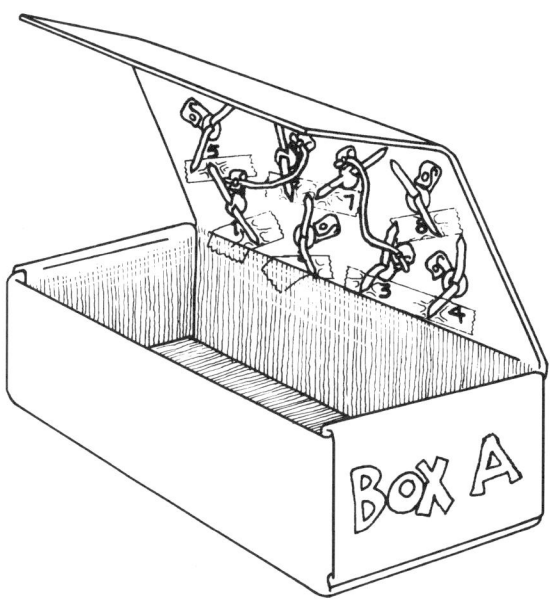

4. Also, to help students keep track of the boxes they have tested, make sure that each box is given a different identification letter.

Procedure

1. Organize students to work in pairs. Have each pair make a circuit tester. Remind them to touch the wire ends to see that it is working.

2. Begin the lesson by saying that you have some puzzles for students to solve. Open the lid of one box and briefly show the wiring to the class. Tell them that each box is wired differently.

3. Demonstrate how to figure out the wiring of a box by following the guidelines listed below:

   ■ Show the class the circuit tester. Ask the students how they would use their circuit tester to find the hidden wires.

- Let one student come to the front of the class to show how to test to find one of the hidden wires. Then ask that student to draw the prediction on the board of which two numbers complete a circuit.

4. Ask the class how they could devise a system that would test every possible pair of contacts on the box. You might ask the class to try first to figure out how many different pairs there are. (There are 28 total pairs. It is an interesting math problem to solve.) Students will need to develop a system to use their time efficiently.

5. Give each pair of students a hidden circuit box. Remind them that you would like them to use their circuit testers to figure out where the hidden circuits are and to draw their findings in their notebooks. The students' drawings, labeled with the letter of the box, will provide you with a record of each student's work.

6. Tell the students that, for the first box only, you want them to draw their predictions, then open the box and see how it is wired. Looking at this first box will give students immediate feedback on their efforts and enable them to modify their technique as needed.

   Remind students that the wiring will look different when viewed from the top of the box lid than it does when viewed from the bottom of the lid.

7. When a pair of students has completed the first box, they should trade boxes with another team.

Final Activities

1. Have students continue to work on the boxes without opening them. About 15 minutes before the end of class, bring the group together to discuss their results. If there is a disagreement, you can open the box to resolve the dispute.

   After a few enthusiastic and successful students have shared their findings, try to draw out those students who are struggling. Use their comments to talk about problem solving and coping with feelings of frustration.

2. Have students return the circuit testers to their own storage boxes and place the hidden circuit boxes in a designated spot in the room.

Extensions

1. Keep several hidden circuit boxes and a circuit tester at the learning center so that students can work on them in their spare time.

2. Suggest to an ambitious team that they design an answer board. (See Figure 9-3.) On the left column, have students devise a list of questions and place a paper fastener to the right of each question. On the right column, have students list the answers in random order and place a paper fastener beside each answer. On the back side of the board, have them connect wires between the corresponding questions and answers.

   Have students use a circuit tester to determine which answer goes with which question. When they connect the wires of their circuit tester to the correct two paper fasteners, the light will go on.

   You can use a large piece of oak tag or poster board for the answer board. Or, you can use a piece of pegboard that already has holes in it.

Hidden Circuits / **55**

LESSON 9

Figure 9-3

Answer board

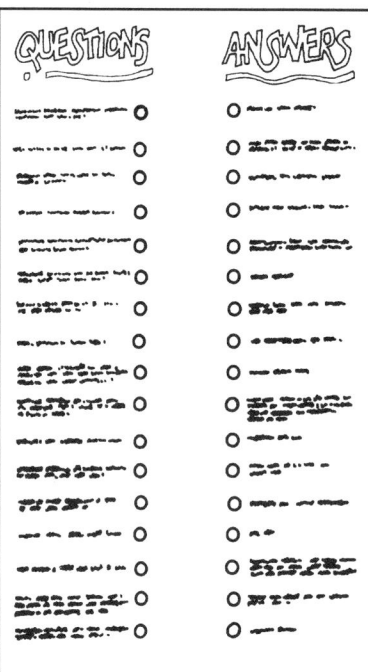
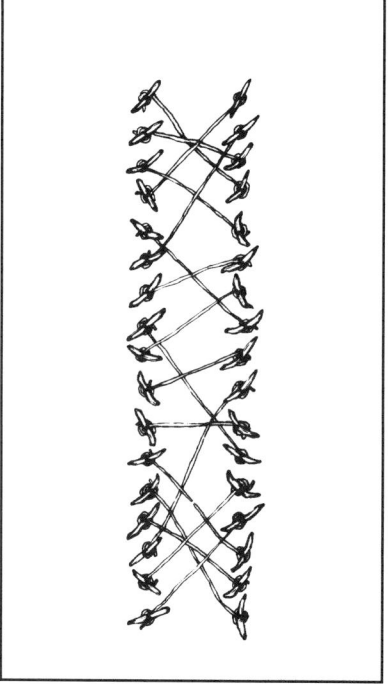

Evaluation

1. Observe students as they work in pairs to find the hidden circuits. Some students will need continued coaching on working together successfully.

2. The drawings in their notebooks will tell you how successful students have been and will reveal any persistent problems. If necessary, bring together a small group for instruction on problem-solving techniques.

LESSON 10

Deciphering a Secret Language

Overview

Circuits in common electric appliances, such as radios, televisions, and refrigerators, are identified by circuit diagrams, which are plans showing how electrical components are connected together. In this lesson, students will learn the symbols used in **circuit diagrams** and will practice making their own circuit diagrams.

Objectives

- Students learn the symbols used in circuit diagrams.
- Students practice translating electrical components into symbols.
- Students practice using circuit diagrams to construct real circuits.

Background

Learning to use symbols makes it easier to draw electric circuits and helps students understand circuit diagrams. Circuit diagrams, plans showing how electrical components are connected, are found on the backs of radios, televisions, refrigerators, and other devices. They contain essential information needed to repair the appliance.

Symbols for the electrical components used in this unit are shown in Figure 10-1. This figure is also placed at the end of this lesson so that it can be made into an overhead transparency.

Figure 10-1

A secret language

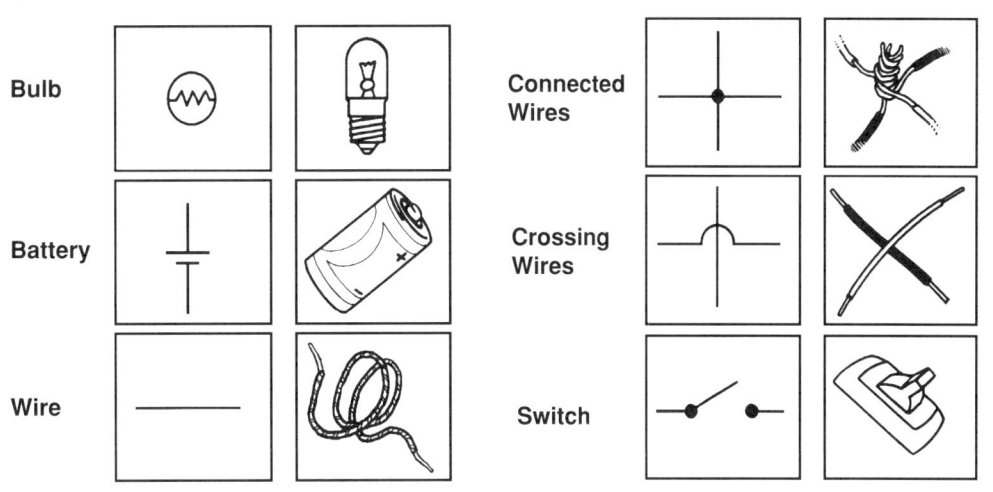

Deciphering a Secret Language / 57

LESSON 10

The circuit diagrams that students draw will not have the neat straight lines and sharp corners seen on commercial circuit diagrams. In the interest of time and brevity, it is best to accept the student drawings as they come. Some students in the 9 to 11 age group are still learning how to translate pictures into symbols. This lesson will given them more opportunities to develop in this area.

Materials

For each student
  1 student notebook

For each pair of students
  1 storage box containing:
      1 D-cell battery
      1 battery holder
      1 bulb
      1 bulb socket
      3 6-inch pieces of wire

For the teacher
  3 D-cell batteries in holders
  3 bulbs in sockets
  6 6-inch pieces of wire

Preparation

1. Familiarize yourself with Figure 10-1, symbols for parts of electric circuits. Draw the symbols on the chalkboard.
2. Prepare one circuit with two batteries in holders, one bulb in a socket, as illustrated in Figure 10-2. A circuit diagram is also shown.

Figure 10-2

Two batteries and a bulb

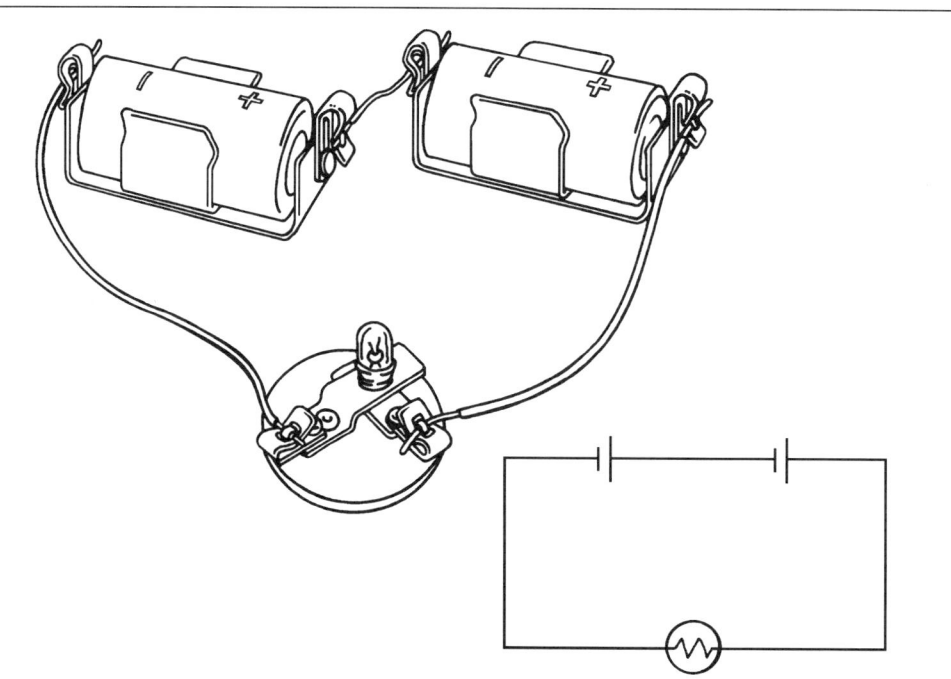

58 / Deciphering a Secret Language

3. Prepare a second circuit with one battery in a holder, two bulbs in sockets, as illustrated in Figure 10-3. A circuit diagram is also shown.

4. Plan to have students working in pairs while drawing circuits.

Figure 10-3

Two bulbs and a battery

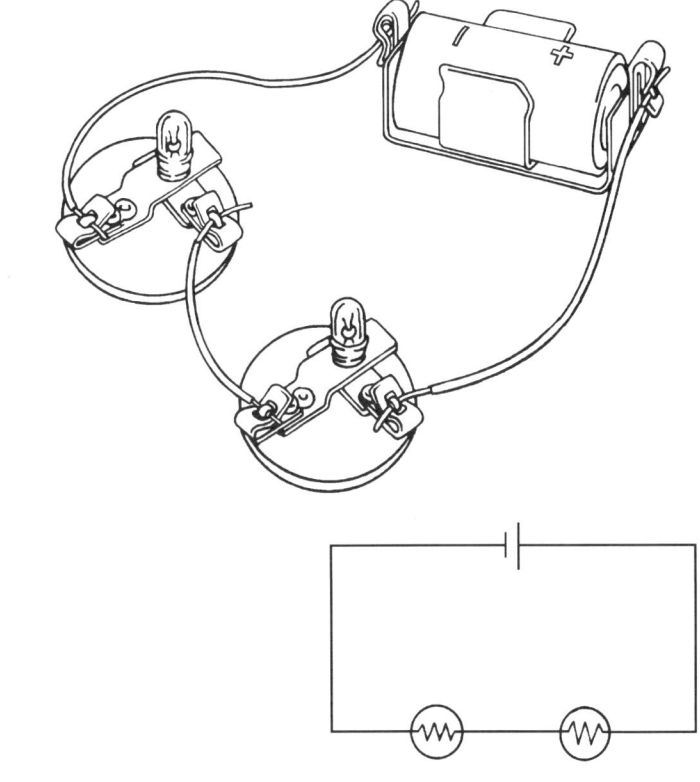

Procedure

1. Tell the class that you are going to teach them a simple way to draw circuits. Tell them that this technique is used by electricians and electrical engineers. Explain that the symbols on the chalkboard are used in circuit diagrams, and that circuit diagrams are the printed sheets often found on the backs of electrical equipment and appliances such as televisions and radios.

2. Display the circuit shown in Figure 10-2 so that everyone can see it. Using the symbols you have already written on the board, draw the circuit diagram for this circuit. Go through it step by step, showing students how you translate from the "real thing" to the drawing.

    As you draw the symbol for the battery, emphasize that the longer line is always the + (positive) end of the battery.

3. Place the circuit illustrated in Figure 10-3 in a place where all students can see it clearly. Tell students you want them to draw a circuit diagram of this circuit in their notebooks. Encourage students to save all their efforts.

4. Tell them you want each member of the team to be responsible for helping his or her partner understand how to use the symbols.

    As students work, circulate around the room. Notice how the pairs are doing. Coach students as needed.

5. Have students return their items to their boxes and the boxes to the designated storage area.

Deciphering a Secret Language / **59**

LESSON 10

Final Activities To end the class, tell students that in the next lesson you will ask them to work together in pairs to make a new circuit. Tell them that they will use the new symbols they have learned to draw the new circuit in their notebooks. To prepare for this activity, have them draw a diagram of one circuit they would like to make next time. Say to them: "Assume you have two batteries and two bulbs to work with. Draw a circuit diagram that includes them in one circuit. Tomorrow you can work with a partner to make that circuit and see how it works."

Evaluation The students' drawings, using the symbols, will provide some insights into each student's understanding of the concepts presented in this lesson.

A Secret Language

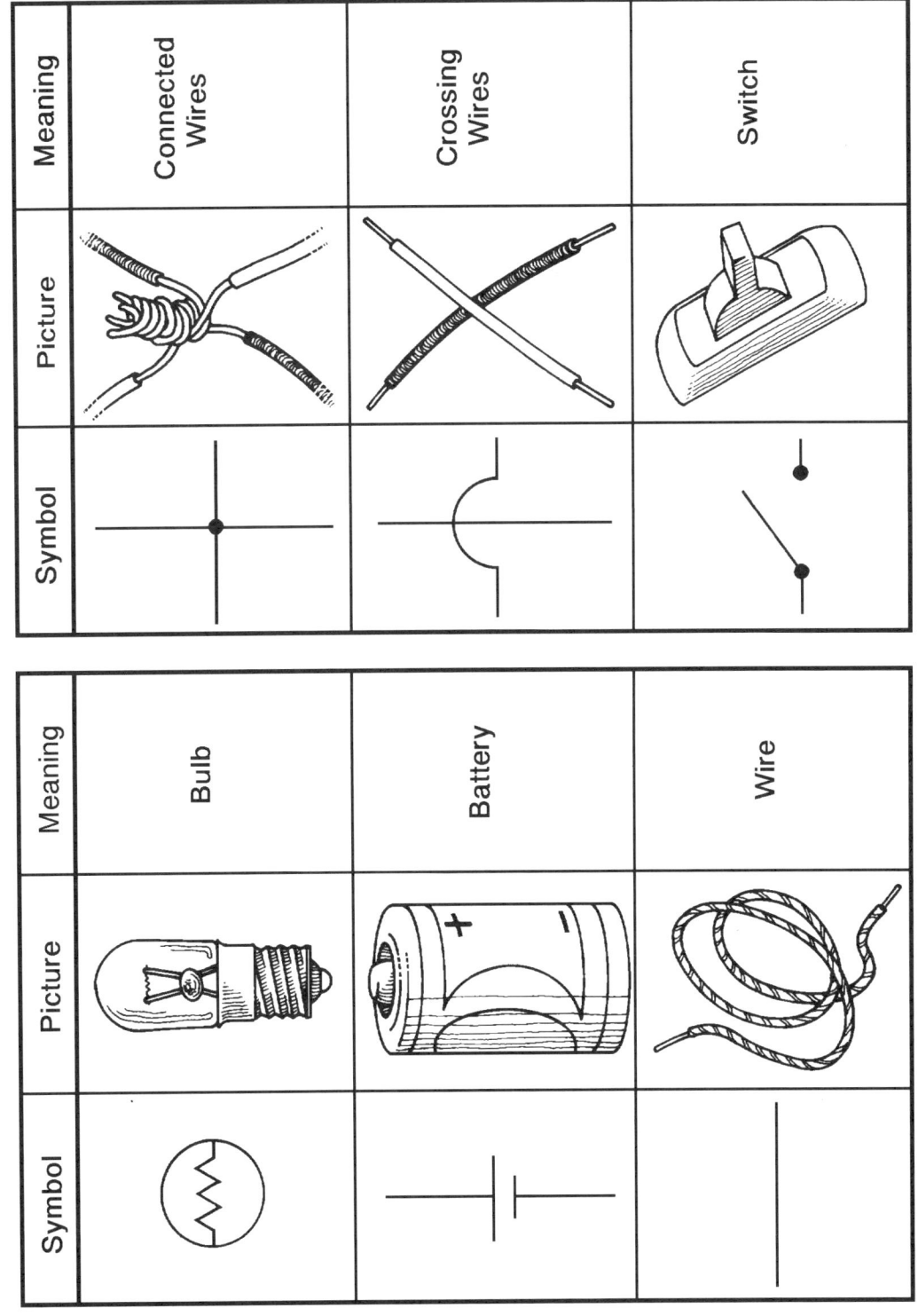

LESSON 11

Exploring Series and Parallel Circuits

Overview

In this lesson, students will build two kinds of circuits: a series circuit and a parallel circuit. They will use the same set of materials for both circuits. By examining the differences between these two circuits, students will learn to identify the properties of series and parallel circuits.

Objectives

- Students build a series and a parallel circuit.
- Students learn to identify series and parallel circuits and begin to use this knowledge to describe their own circuits.

Background

There are two kinds of circuits: series circuits and parallel circuits. In Figure 11-1, the batteries are arranged in series. (The circuit diagram has been included.) In a series circuit, electricity has only one path to travel from one point on the circuit through the wires, batteries, and bulbs and back to the starting point.

Figure 11-1

Batteries in series

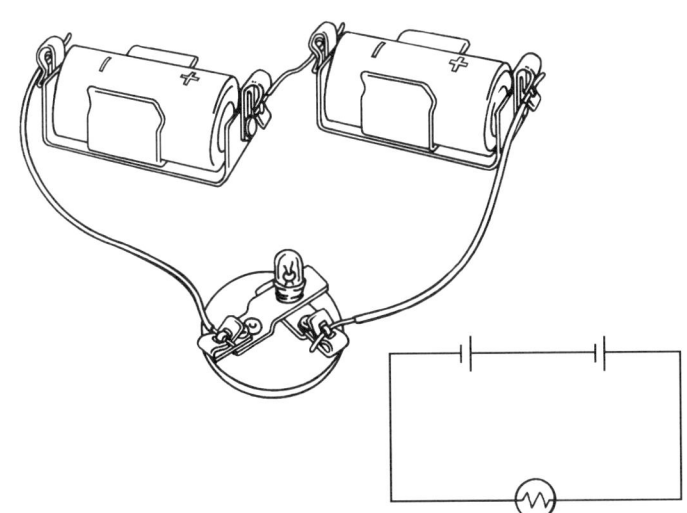

Exploring Series and Parallel Circuits / **63**

When batteries are arranged in series, the voltage across the bulb is increased, causing the bulb to glow brighter than it did when only one battery was used. But the batteries will drain more quickly.

In Figure 11-2, the batteries are arranged in parallel. In a parallel circuit, electricity travels along more than one path around the circuit. When batteries are arranged in parallel, the brightness of the bulb will be the same as it was with one battery. But the bulb will burn longer in this circuit than it will when the batteries are arranged in series.

Figure 11-2

Batteries in parallel

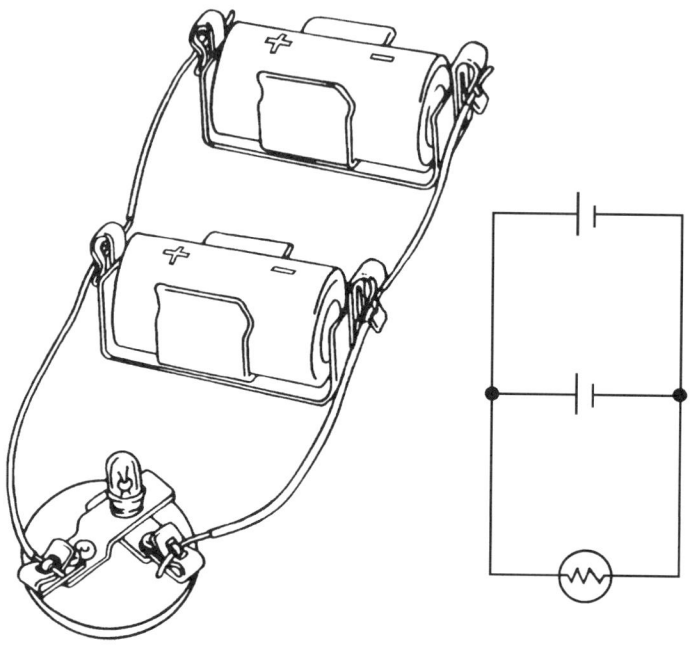

Bulbs can be wired in series or in parallel, too. Figure 11-3 shows bulbs arranged in series. When two identical bulbs are wired in series with one battery, they burn with uniform brightness, but they are not as bright as one bulb alone.

Figure 11-4 shows bulbs arranged in parallel. When two bulbs are wired in parallel with one battery, each bulb burns as brightly as a one bulb/one battery arrangement. One attribute of the parallel circuit of bulbs is that unscrewing one bulb does not make the other bulb go out. The electricity travels in independent paths through each bulb.

Appendix F includes a more detailed description of series and parallel circuits. This is for your information and is probably beyond the scope of what students can assimilate now. When working with students, encourage them to observe what happens when bulbs and batteries are put in different circuit arrangements. Focus on their observations and documentation of the phenomena. Explanations can come in later years.

Figure 11-3

Bulbs in series

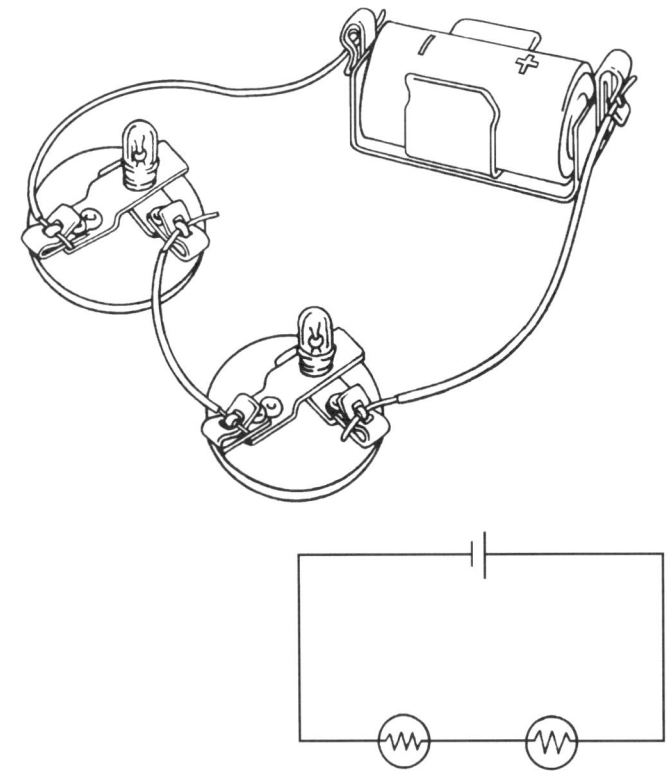

Figure 11-4

Bulbs in parallel

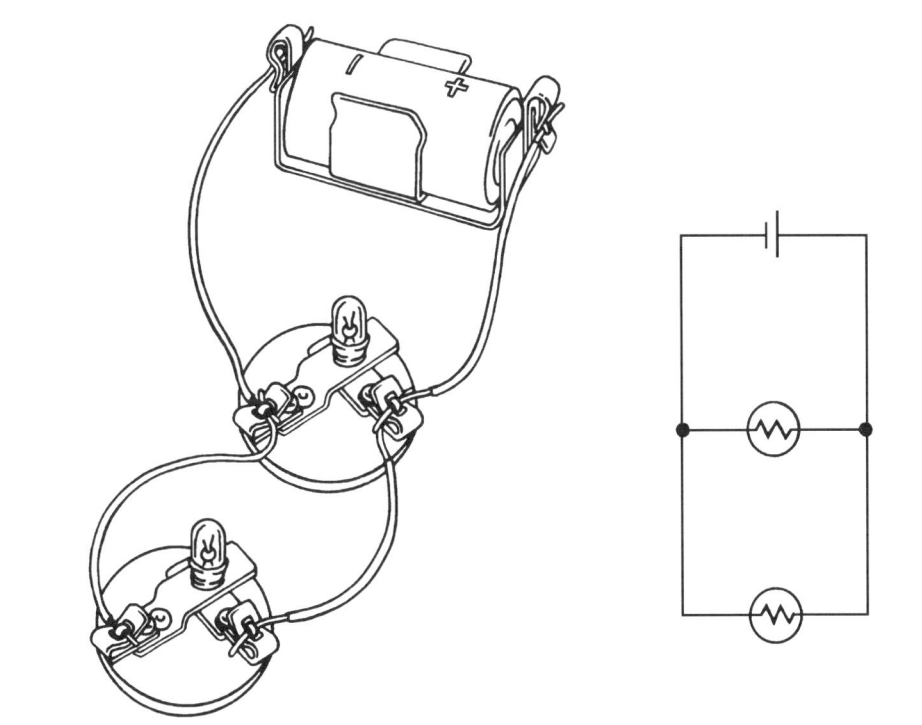

Exploring Series and Parallel Circuits / **65**

LESSON 11

Materials

For each student
1 student notebook

For each pair of students
2 storage boxes, each containing:
    1 D-cell battery
    1 battery holder
    1 bulb
    1 bulb socket
    3 6-inch pieces of wire

Procedure

1. Have students work in pairs. Make sure each pair has two bulbs and two batteries. Emphasize to students the importance of discussing the activities with each other before proceeding.

   At the end of Lesson 10, some students drew a circuit diagram of another circuit they would like to build. You might give them an opportunity to construct that circuit first.

2. Draw on the board the circuit diagrams from Figures 11-1 and 11-2. These circuit diagrams are also in the Student Activity Book. Ask students to predict which of the bulbs they think will be the brightest and which they think will burn for the longest time.

3. Ask students to use each circuit diagram to actually put together the circuit. Each pair of students will have to make a decision about which circuit to make first.

4. When the pairs have constructed each circuit, have students record the results in their notebooks. In particular, have them keep track of the brightness of the bulbs in each case.

5. Help students develop a way to judge the degree of brightness of the bulb. The class can decide that a simple one bulb, one battery brightness is the standard. Students can then judge another bulb's brightness by comparing it with "the standard" bulb, classifying it as "dimmer," "about the same," or "brighter" than the standard bulb.

Final Activities

1. Discuss the circuits in Figures 11-1 and 11-2. Ask the class to look at the differences between the two circuits, focusing on the brightness of the bulbs. Compare both bulbs to the "standard" one battery brightness. Tell the students that in the first circuit, the batteries are placed in series. In the second circuit, the batteries are in parallel.

2. Discuss the terms "series" and "parallel." Ask students to tell what they think they mean. Finally, ask them to write each term in their notebooks and draw a circuit diagram illustrating each type—this time with bulbs in parallel and bulbs in series.

3. Place the equipment in the storage boxes.

Extensions

1. Put on the chalkboard the circuit diagrams for Figures 11-3 and 11-4. Ask the students to make these circuits to see what happens when the two bulbs are in parrallel, compared with when the two bulbs are in

series. Have them draw the circuits in their notebooks and record the brightness of the bulbs.

2. Set up an experiment to see which circuit will burn longer: the series circuit in Figure 11-1, the parallel circuit in Figure 11-2, or the "standard circuit" of one bulb, one battery. It would be best to set up this experiment on a morning early in the week so that students can check the circuit at the beginning and at the end of each day.

    Ask students to draw a diagram of all those circuits in their notebooks and to predict which bulb they think will burn the longest. These circuits can go for many days, depending on the strength of the batteries.

3. After all three bulbs have gotten dim enough that all agree they are "burned out," discuss the results. Ask students what they think caused the bulbs to go out. Give them time to think, and accept all of their ideas for discussion. Some students will think the bulb burned out and some will think the batteries drained. Some will have ideas you would never imagine!

    After the class has suggested some reasons, ask: "How could we check these ideas?" Use this as an opportunity to use the troubleshooting techniques discussed in Lesson 6 to find out whether the bulb or the battery was responsible for the light going out. Ask for volunteers to check the batteries and the bulbs.

Evaluation

Students' drawings and predictions will provide information about their understanding up to this point.

| LESSON 12 | **Learning about Switches** |

Overview

Many common electrical devices use both parallel and series circuits. A flashlight, for example, uses a series circuit for maximum brightness. In this lesson, students think about these concepts by drawing a plan for a flashlight. In the next lesson, they will construct a flashlight. To make the flashlight, students need to learn about another important element of complex circuits: the switch. These lessons help students see how all the parts of a circuit work together

Objectives

- Students construct a switch and learn why switches are important.
- Students apply what they have learned about series and parallel circuits to devise a plan for a flashlight.

Background

Flashlights come in many shapes and sizes, but they have several things in common. They have a power supply (the batteries), a light bulb, wires or pieces of metal to complete the circuit, and a switch. The switch, an important part of the circuit, turns the flashlight on and off. In addition, most flashlights have a lens or mirror to focus the light in a particular direction.

Materials

For each student
- 1 storage box containing:
    - 1 D-cell battery
    - 1 battery holder
    - 1 bulb
    - 1 bulb socket
    - 3 6-inch pieces of wire
- 1 student notebook

For every two students
- 2 6-inch pieces of wire
- 1 paper clip
- 1 3" x 5" index card
- 2 No. 3 brass paper fasteners

LESSON 12

2 brass paper fastener washers
2 Fahnestock clips

For the class
1 roll ¾" masking tape

Preparation

1. Before class, make a simple switch according to the plan shown in Figure 12-1. This drawing also shows the symbol for a switch used in circuit diagrams. Students will need to use this symbol in their diagrams.

Figure 12-1

Paper clip switch and the symbol for a switch

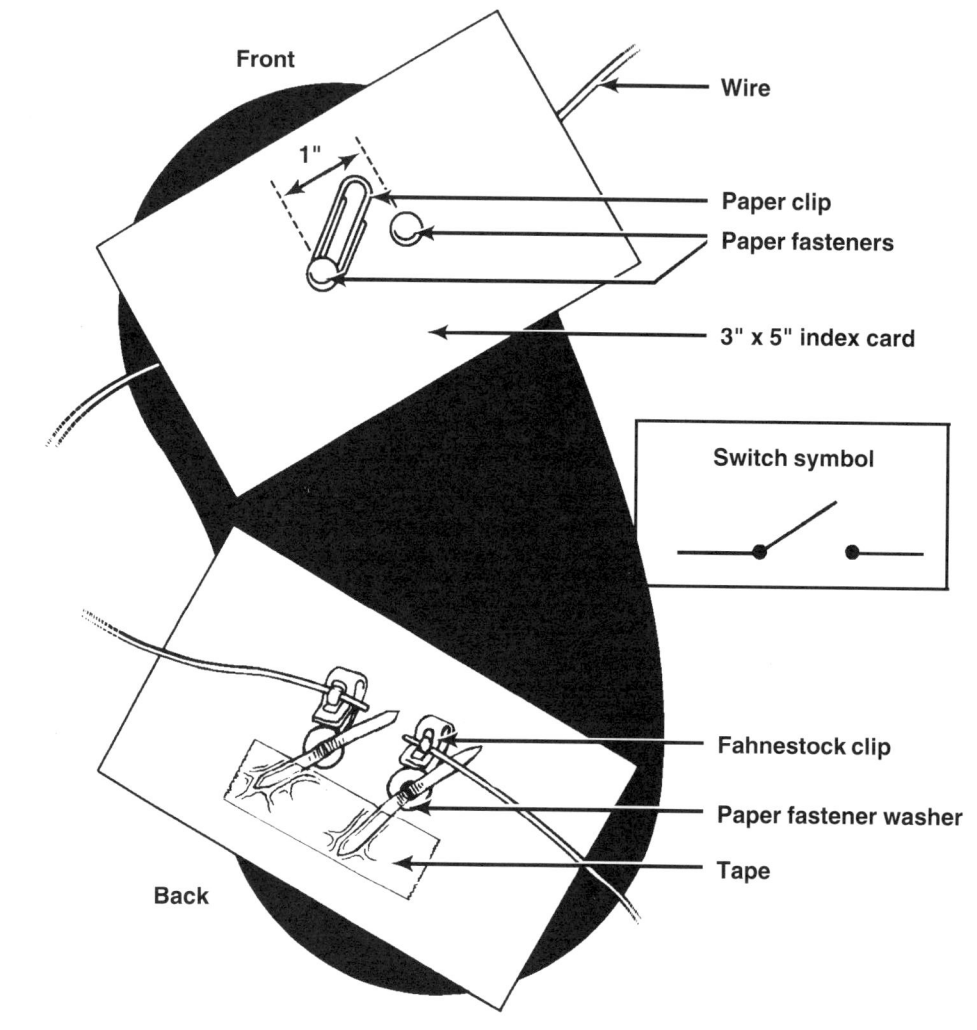

Note: The paper fasteners should not touch each other on the back side of the card. Use a piece of masking tape, as shown in Figure 12-1, to hold the ends of the paper fasteners apart.

2. Put the switch in a circuit that has one bulb and one battery, all in series, as illustrated in Figure 12-2. Use the switch to turn the light on and off.

Figure 12-2

Switch in a series circuit and its circuit diagram

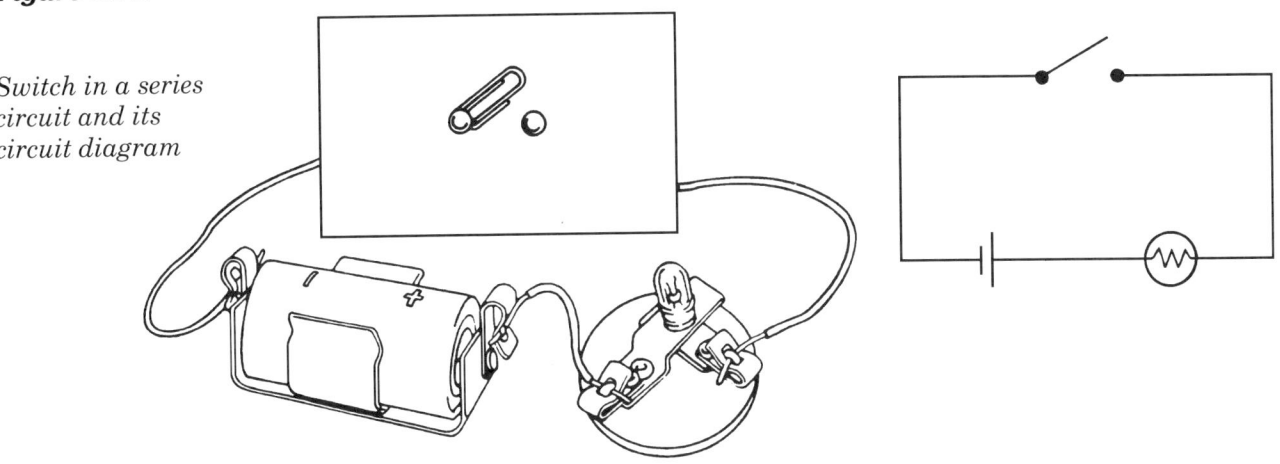

Procedure

1. Start the class by telling the students that today's task is to come up with a plan for a flashlight. Ask students what they think are the important parts of a flashlight. As the students make contributions, write them on the chalkboard or on chart paper in one of two columns: "Necessary" and "Useful."

2. After a few minutes of discussion, the items in the "Necessary" column should include:

   - a power source (the battery)
   - a light (the bulb)
   - wires or conductors
   - a way to turn the flashlight off and on (switch)
   - that it be portable

   Save this list to use in Lesson 13. Use the "Useful" column for comments such as "It is made of hard metal or plastic," or "It is waterproof." These are items that the students can ignore or, if they wish, incorporate in their invention.

3. Now that students have had a chance to think about the components of a flashlight, tell them that they will have a chance to build their own. Tell each pair that before they can construct a flashlight, they will need to figure out a way to turn the light on and off. To give them a starting point, show them the paper clip switch you have made. Demonstrate how to use it to turn the light on and off.

4. Invite the students to make their own switch. They could make one like the one you demonstrated. Or, you could give them the opportunity to invent their own. (See item #3 in **Extensions**.) The Student Activity Book includes plans and instructions for making the switch shown here in Figure 12-1.

5. Circulate around the class and make sure that each pair of students can successfully make a switch and use it to control a simple circuit, like the one in Figure 12-2. Make sure that the switch turns the light on and off.

Learning about Switches / 71

LESSON 12

6. When each pair has made a switch and wired it into a circuit, ask students to draw in their notebooks a circuit diagram that includes the switch.

7. Ask students what the basic positions of the switch are. The class will agree that there are two: "on" and "off." Share with the students the convention that a switch that is "off" is referred to as "open" and a switch that is "on" is referred to as "closed." (These words refer to the fact that the circuit is "open" or "closed.")

8. About 10 minutes before the period is over, have students put their switches in their storage boxes and clean up. Tell them they will use the paper clip switch in the next lesson, when they will make a flashlight.

Final Activities

Ask each pair of students to use the next 5 minutes to plan the flashlight they will invent. Remind them of the essential elements of the flashlight, and ask them to plan their flashlight using two batteries and one bulb. Their plan could result in a picture or in a circuit diagram, or both, drawn in their notebooks.

Extensions

1. Ask students to think about where they have seen switches. On the chalkboard, draw the symbol for the switch. Under the symbol, list kinds of switches that students mention. (Some examples are the ignition switch on a car, the light switch on a car, the light switch on a wall, and the on/off switch on a hair dryer.) This will strengthen the association between different switches and the symbol that represents them in a circuit diagram.

2. Have students go on a "switch" scavenger hunt. How many kinds of switches can they find? Ask them to make a list to share with the class.

3. Invite students to invent a different switch. In addition to the materials listed, provide aluminum foil, wooden clothespins, and a stapler.

Evaluation

1. As students work to make the switch and wire it into a circuit, you will see how well they are able to apply what they are learning.

2. The circuit diagrams students draw in their notebooks provide information about each student's understanding of circuits and how they work. If the drawings have the essential elements—a battery, a bulb, a switch, and wires—in a circuit, you will know that the students know the parts of a circuit.

| LESSON 13 | **Constructing a Flashlight** |

Overview Building on the groundwork laid in Lesson 12, students incorporate the switch they have made into a flashlight they have designed. As students share the flashlights they have made, they also review series and parallel circuits.

Objectives
- Students construct flashlights.
- Students discuss the similarities and differences between series and parallel circuits.

Background During this lesson, students have an opportunity to apply what they learned in Lesson 11 about series and parallel circuits to the construction of a familiar electrical device—the flashlight.

If students use a series circuit in their flashlight, the bulb will be brighter. If students use a parallel circuit in their flashlight, the bulb will last longer. As students share their flashlight constructions, these differences between the two circuits will become clear.

Materials *For each student*
  1 student notebook

For every two students
  2 storage boxes, each containing:
      1 D-cell battery
      1 battery holder
      1 bulb
      1 bulb socket
      3 6-inch pieces of wire
  1 paper clip switch

For every four students
  4 sheets of 8½" x 11" construction paper
  1 pair of scissors

LESSON 13

1 roll ¾" masking tape
Crayons or markers

Preparation

Put on the chalkboard the list of essential properties of a flashlight prepared during the previous lesson.

Procedure

1. Distribute the materials and review with students the essential components of a flashlight.

2. Tell them that now they will be given time to build their own flashlights. Have students use the designs they made in Lesson 12. If they didn't complete their designs, have them use one from a classmate or build a flashlight without a design.

3. Have student pairs grouped so that two pairs can share one pair of scissors, one roll of masking tape, and several sheets of paper.

4. As students construct their flashlights, make sure they are working together and have the materials they need.

5. When students finish their flashlights, they will be quite excited and will undoubtedly want to show them off. As students finish, ask them to share with you how the flashlight is turned on and off.

6. Have students make drawings or circuit diagrams of their flashlights in their notebooks. Some students will only be able to make a drawing.

7. Ask students who finish early to spend some time decorating their flashlights.

8. About 5 or 10 minutes before the end of the period, have students clean up. Make sure they leave their flashlights intact, but have them put away all other materials.

Final Activities

1. Ask students to share with the class the flashlights they have created. This could be an opportunity for an oral presentation by the different pairs. Their presentation could include a description of the process of building flashlights, any difficulties overcome, a demonstration of how the flashlight works, and a large version of the circuit diagram drawn on the board.

2. Ask students to think about what kind of circuits their flashlights use. Do they use a series circuit or a parallel circuit?

3. Leave the flashlights on a table for students to look at until the next lesson. At that time, students will need to disassemble them to use their materials for the remaining lessons. But by leaving them on display for even a short while, students will have an easier time taking them apart than if they were asked to do it immediately after building them.

4. You may wish to keep a few of the flashlights on hand for an end-of-the-unit demonstration for parents. If you choose to save a few of them, plan to give those students who are not disassembling their flashlights the extra materials they will need for future lessons.

   If you have a camera available, you might take photographs of the flashlights or of students with their flashlights.

Extensions

1. Students may need extra time to finish their inventions. Allow them to work on their flashlights in their spare time.

2. Ask students to draw and make a circuit diagram of a flashlight they have at home. Have students bring the diagrams in and demonstrate the way different lights use different circuits and different batteries.

Evaluation

The flashlights that students constructed reveal what students understand. Here are some specific things to look for:

- Is the student able to construct the switch and put it in the circuit so that it turns the light on and off?
- Can the student wire the circuit so that the batteries are in series or in parallel?
- Can the student draw a wiring diagram of the flashlight using the correct symbols for the parts of the circuit?

LESSON 14

Working with a Diode

Overview

In many common electronic devices, such as battery chargers, radios, and televisions, it is important that electricity flow in only one direction. A device that makes this possible is called a **semiconductor diode**. In this lesson, students work with a semiconductor diode and learn that it conducts current in only one direction. They also learn that semiconductors are used widely in computers and other electronic devices.

Objectives

- Students experiment with semiconductor diodes and learn how they work.
- Students learn the symbol for semiconductor diodes used in circuit diagrams.
- Students discover the relationship between the passage of current through the diode and the positive and negative terminals of the battery.

Background

The semiconductor diode consists of a small cylindrical piece with a wire coming out of each of the two ends. The cylindrical piece is the functional part of the diode; the wires are just connectors.

The diode is a device that allows current to flow in one direction, but not in the other. This is such a simple and a unique property that the significance of it is often lost. But this is in marked contrast to the wire students are using, and to all the items they have tested and found to be conductors. If you check, you will find that they conduct electricity in both directions.

Figure 14-1 indicates the direction the current flows and the way the symbol for the diode corresponds with that direction. Note that the colored band on one end of the diode corresponds to the direction of current flow through the diode.

Figure 14-1

Diode and its symbol

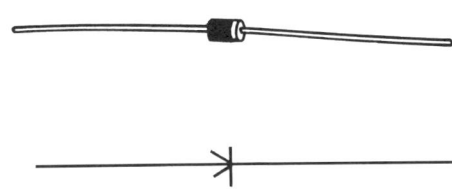

Working with a Diode / **77**

LESSON 14

Materials

For each student
    1 student notebook

For every two students
    2 storage boxes, each containing:
        1 D-cell battery
        1 battery holder
        1 bulb
        1 bulb socket
        3 6-inch pieces of wire
    1 simple diode
    2 Fahnestock clips
    1 paper clip switch

Preparation

1. Before class, take some time to work with the diode so you can see how it functions. Using a circuit tester, touch the ends of the wires to the ends of the simple diode. Does the bulb light? Turn the diode around and repeat the procedure. Does the bulb light now?

2. Use two Fahnestock clips to connect the wires to the diode, as shown in Figure 14-2.

Figure 14-2

Diode in a circuit and the circuit diagram

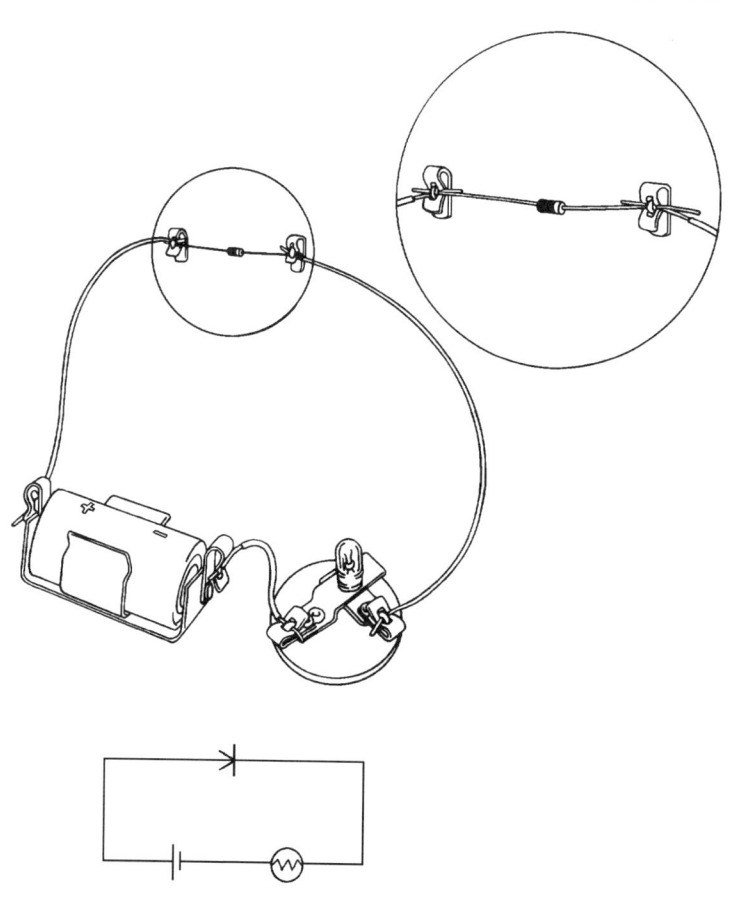

Procedure

1. Still working in pairs, have students begin by taking apart their flashlights so they can use the materials in this and subsequent lessons.

2. Have each pair of students make a circuit tester, but this time have them use two batteries in series and one bulb. Ask them to demonstrate that the circuit works by touching the end wires together.

3. Give each pair a simple diode. Ask the pair to test it with their circuit tester by touching the wires to each end of the diode. For some students, the bulb will light, and for others it will not. Ask the class: "Look closely at the diode. What do you think might be the reason that some bulbs light and others do not?" How diodes work is a fascinating and complex subject. What is more understandable to the students at this point is the fact that the diode's direction makes a difference. You may want to suggest that some students reverse it—and see what happens.

4. Challenge students to put the diode and the simple paper clip switch (used in making the flashlight) into the circuit using two batteries and one bulb. Can they connect it so the switch turns the light on and off?

5. Introduce the symbol used to draw a diode. Ask students to draw in their notebooks a circuit diagram that includes their circuit tester, the switch, and the diode.

Final Activities

1. Ask two students to draw their circuit diagrams on the chalkboard. Have students draw them so that the batteries are pointing in the same direction on the board as they are in their circuits.

    Ask students what the two circuits have in common. If no one sees it, point out that the diode is always pointing in the same direction in relation to the battery when the bulb lights and that the direction of the diode is determined by its circular marking.

2. Emphasize that diodes will pass electricity in only one direction, in the same way a one-way street controls traffic. Contrast this with the behavior of the wire you have used. It passes electricity in both directions.

3. Ask students where they might find diodes. If they don't know, explain that the diode is one of many elements used in the circuits of radios, televisions, and other electronic devices.

Extensions

1. *Introduction to Electronics* by Pam Beasant is one of many books about electronics that offers those who are interested and motivated more information about the subject. (See **Appendix G**.)

2. If some students show an interest in doing more experiments with electronics, there are numerous kits available in electronics stores that enable students to make simple devices and learn more about electronics.

Evaluation

If students are able to make the two-battery circuit tester, put the diode in the circuit, add the switch to the circuit, and then draw a circuit diagram of the arrangement, they are demonstrating a high level of competence with this material.

Working with a Diode / **79**

LESSON 15

Planning to Wire a House

Overview

At this point in the unit, students have learned many important concepts related to electricity, including how to make circuits, the difference between series and parallel circuits, and how to make a switch. In this lesson, students will have a chance to apply what they have learned by drawing up a plan for wiring a cardboard box "house." In the final lesson, students will implement their plans.

Objectives

- Students work in teams to use knowledge gained during the unit to draw up plans for wiring a house.
- Students consider different strategies for making an effective wiring scheme.

Background

There is some similarity between the wiring students do to light a cardboard box and the wiring done in a house. For example, students will need to grapple with the way the wires go from the power source (batteries outside the box), through the walls and then into each room, just as electricians do. They will also need to plan for the placement of the switch and the light in a realistic way, with wires going between them.

Students have many other issues to think about as they work on this project. For example, should they use only one D-cell and one light for each room, which would keep the wiring diagram quite simple and self-contained within each of the four rooms? Or, should students use two D-cells as a power source, and have them placed "outside" the house, similar to the way power is sent from the electric company to "real" houses? Encourage students to consider both alternatives.

Another thing students should consider is how bright to make their houses. You can challenge the students to make the lights as bright as possible, using the two D-cells for power. (To make the lights particularly bright, students should wire the D-cells in series and the bulbs in parallel.)

The students also can use paper clip switches to turn the lights in their house on and off. The placement of the switches can offer other challenges to the students. For example, at the simplest level the students might use only one switch to control all the lights. At a more complex level, they might place a switch in each room to control only the light in that room.

LESSON 15

Materials

For each student
1. student notebook

For every four students:
1. cardboard box
2. sheets of 8½" x 11" drawing paper
1. pencil
1. colored pencil or crayon

Preparation

1. Obtain one cardboard box for every four students. The box should be approximately 12" x 12" x 18". A box that holds duplicating paper or a box from a grocery store will work well.

    Keep the dividers that sometimes come in the boxes. Use the dividers to make four little "rooms" in each box, as illustrated in Figure 15-1.

Figure 15-1

Using dividers in a cardboard house

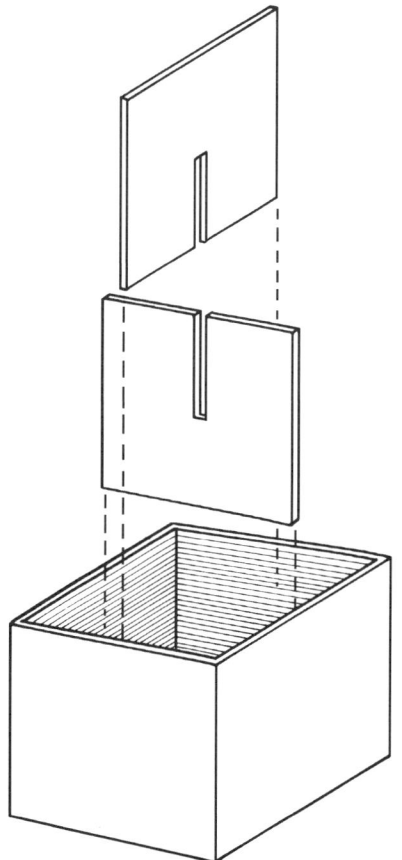

2. Plan to have students display the wired boxes as part of an exhibit for parents or for other classes after you have completed the unit.

82 / Planning to Wire a House

Procedure

1. Tell students that they will be wiring the box as though it were a house. Organize them in teams of four. Have each team choose one person to be the discussion leader and another to do the drawing. Have each team work together to draw a plan for their house wiring.

2. Speaking to the class as a whole, remind the students of the work they have already done designing circuits, looking at the differences between series and parallel circuits, and using switches. Discuss how using series and parallel circuits can affect the brightness of the bulbs.

3. Tell students that their task is to organize their house so that there is a light in each room. Draw on the board the simple pattern of the four rooms of the house. Say to them: "When we make a plan for the house, what do we need to decide?"

   As students discuss this, write essential questions on the board. Some of those questions are listed below. Feel free to add any others you or the students think are important.

   - Where will the lights be placed?
   - Where will the switches be placed?
   - Where will the D-cell batteries be placed?
   - How should the switches, bulbs, and D-cell batteries be connected so that the bulbs burn the brightest?
   - Where will the wires be placed?

4. Remind students that their plans will be easier to read if they use the symbols for the elements of the electric circuits (see Lesson 10). It will also be easier to read the plans if they use a different color pen or pencil for the cardboard walls and for the wiring.

5. Have students go to work. Circulate around the room, watching and serving as a resource person where needed. By the end of the period, each group should have some version of a plan for the wiring they will do in the next lesson.

6. Ask students to draw a copy of their wiring plans in their notebooks. These drawings will give them a record and will give you an opportunity to see how each student is doing.

7. Clean up.

Final Activities

When most students have finished, ask them to share their plans. This will provide an opportunity for an oral presentation, and it will help students who are struggling with their plans to see what others are doing.

Extensions

Some students will benefit from extra time to perfect their plans. Between Lessons 15 and 16, those students should continue working on their plans so they are ready to go to work wiring at the beginning of the next class.

Evaluation

The house plans will provide information about how well students use symbols in their wiring diagrams and about their understanding of circuits.

Planning to Wire a House / **83**

LESSON 16

Wiring and Lighting the House

Overview
For the unit's final lesson, students use their plans to wire and light the cardboard box house. By the end of the lesson, students will have learned some important information about how a real house is wired.

Objectives
- Students apply what they have learned about series and parallel circuits.
- Students use all the skills and information they have gained to work in teams to wire and light a house.

Background
The plans students have made will vary in quality. Expect that students will use those plans only as a way to begin. As they work, some students will see ways that the plans can be improved. Other students may have difficulty translating their plans into the real wiring of their model house.

In any event, modifications will be made and should, to some extent, be encouraged. In the real world, builders frequently see new possibilities and problems as they implement plans, and they, too, make adjustments and changes as they work. Encourage students to make a new wiring diagram showing what they actually did after they have completed the project.

Materials
For each student
- 1 student notebook

For every four students
- 1 cardboard box
- 4 storage boxes, each containing:
    - 1 D-cell battery
    - 1 battery holder
    - 1 bulb
    - 1 bulb socket
    - 4 6-inch pieces of wire
- 1 roll of 1" masking tape
- 2 paper clip switches
- 1 pair of scissors

LESSON 16

For the class

- 1 roll of wire
- 1 wire stripper
- 2 screwdrivers
- 30 3" x 5" index cards
- 1 box of paper clips
- 1 box of No. 3 brass paper fasteners
- 1 box of brass paper fastener washers
- Crayons, paints, or markers

Preparation

1. Identify any student groups who have not developed a plan. To help them get started you might either:
   - have them use the plan from another group
   - have them wire the house first and draw the plan afterward

2. While wiring the house, students will need extra wire, the materials to make extra switches, and screwdrivers. Decide how you want to dispense these materials. If you want to keep control of things, hand out the materials yourself. If you have an extra helper in the room, consider putting that person in charge of dispensing materials. This will free you up to move around the room, observing and helping out as needed.

3. Some groups will want to decorate their houses. They will need crayons, paints, or markers.

Procedure

1. Remind students of the excellent plans they have created or have available for their use. Tell them how you will handle any needs for more wire or other materials. Urge them to keep working together and to talk to members of their group as they work so that each member has an opportunity to contribute.

2. Point out that the screwdriver can be used to punch a hole in the cardboard to allow wires to pass through the "wall" of the house. Warn students to be careful when using it. The masking tape can be used to hold things as needed, such as wires along a wall and the switch to the wall. Have materials available for students to make additional switches.

3. Encourage students to talk to each other as they work. Remind them to modify their plans, if necessary.

4. As students finish, ask them to make a wiring diagram of the actual wiring in the house and to put it in their notebooks. Have them compare their actual plan with their original design.

5. When students have finished wiring and lighting, have them decorate their houses.

6. Clean up.

Final Activities

Have each student group present their house to the class with an oral presentation and a demonstration. Or, have each group prepare a written description of their house. The written description could recount, among other things, the work they did in making their plans, what issues and

86 / Wiring and Lighting the House

Extensions

1. Here is a challenge for those who enjoy a complex problem. Consider a room with two doors and one light. Put a switch by each door and wire the room so that either switch can be used to turn the light on *or* off.

   To accomplish this, the students will need switches like the ones shown below. These are called "single-pole, double-throw" switches. This kind of switch is illustrated below.

Figure 16-1

Single-pole, double-throw switch and its symbol

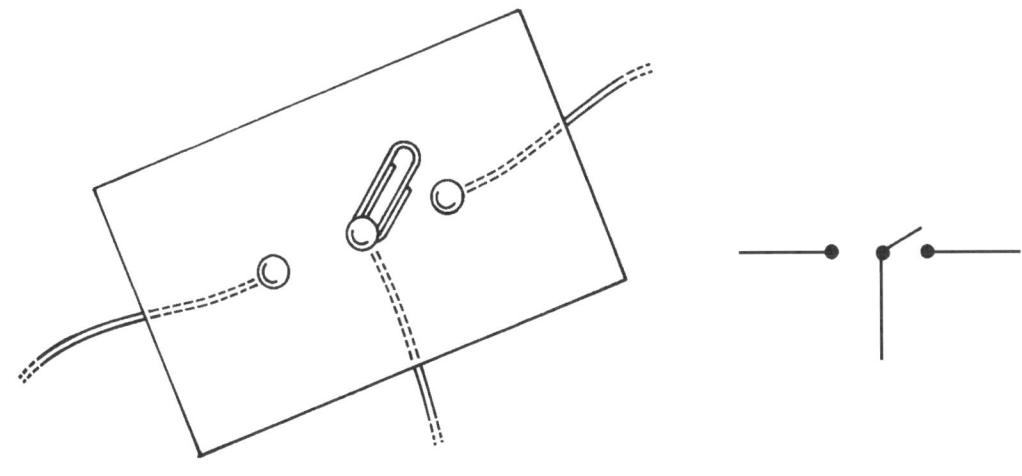

2. Have students write a story describing how they wired their houses. They might recall interesting or amusing examples of problems they had and how they resolved them.

3. Invite an electrician to come speak to the class. He might describe some of the challenges faced in wiring houses or buildings, as well as other aspects of his job. An alternate speaker might be a representative from the local power company.

Evaluation

1. The task of wiring the house offers an excellent way to evaluate what students have learned up to this point. You may wish to take notes as you observe students working on this project.

2. The house itself can be decorated and used in a presentation to other classes or to parents. See **Appendix A** for suggestions for post-unit assessments.

Wiring and Lighting the House / **87**

APPENDIX A

Post-Unit Assessments

Overview
- **Assessment 1** is a follow-up to the student brainstorming session about electricity held in Lesson 1.
- **Assessment 2** offers suggestions for displays of student products.
- **Assessment 3** offers suggestions for evaluation of student products.
- **Assessment 4** consists of suggestions for student experiments.
- **Assessment 5** is a paper-and-pencil test for students.

Objectives
- Students evaluate their own progress.
- The teacher evaluates student progress.

Materials
Since materials will vary, they are listed separately at the beginning of each assessment.

ASSESSMENT 1

A Follow-up to the Student Brainstorming Session about Electricity Held in Lesson 1

During the brainstorming session in Lesson 1, students made a list of questions they had about electricity. Now students have an opportunity to discover how much they have learned.

Materials
The list saved from Lesson 1.

Procedure
1. Display the list from Lesson 1. Discuss the following points:
   - Ask students to identify statements on the list that they now know to be true. What experiences did they have during the project that confirmed these statements?

APPENDIX A

- Ask students to identify statements that need correction or improvement. What corrections/improvements do they want to make? Based on what?

2. Ask students to read their notebooks and look at their drawings. Ask them to write a final entry describing what they now know about electricity.

3. Students drew a lightbulb at the beginning of Lesson 2, and then again at the end of Lesson 4. The comparison of these two drawings demonstrates one thing students have learned.

ASSESSMENT 2   Student Display

Students create a display for other classrooms or for parents.

Materials   Student products.

Procedure

1. Here are some suggestions of items that would make attractive displays:
   - The houses, wired and lit, with wiring diagrams displayed.
   - Some of the flashlights students invented, with wiring diagrams.
   - Regular commercial flashlights for which students have drawn a wiring diagram and written an explanation of the way they work.
   - Different circuits, to illustrate the differences between series and parallel circuits.
   - A display of the symbols used to make circuit diagrams. The display could include some of the students' diagrams, an explanation of the symbols, and some circuit diagrams that students have gathered from home appliances.
   - Student drawings that show the parts of a light bulb.
   - Boxes with hidden circuits for visitors to try to solve.
   - Answer boards that the class created.
   - A demonstration of the diode.
   - A display reflecting books that were read. Perhaps a reading or a dramatic presentation from the lives of Franklin or Edison might be enjoyable.

2. The project in Lesson 16 is an assessment tool in itself. The plans for wiring the house and the wired box provide direct evidence of what the students understand and are able to do. Your observation of the way each group functioned also gives you evidence about their ability to work as part of a team.

3. Student understanding of the project also is reflected in the final wiring diagram of the house drawn for student notebooks.

APPENDIX A

ASSESSMENT 3 **Student Portfolio**

The teacher uses students' work to evaluate their progess.

Materials Student products.

Procedure
1. Gather together all the work that students have done. Include the following items:
   - student notebooks
   - all activity sheets
   - all student writing on science
   - any book reports on science topics
   - student drawings
   - student projects and inventions. This could include the actual project, an invention, or a description and a picture of the project.
2. Use those products to assess how much students have learned.

ASSESSMENT 4 **Student Experiments**

The following experiments are an alternative way to assess student understanding of key concepts.

Materials See individual experiments.

Procedure *Experiment 1*
1. Give students the following materials, disconnected:
   - 1 battery
   - 1 battery holder
   - 2 bulbs
   - 2 bulb sockets
   - 4 6-inch pieces of wire
2. Instruct students to construct a circuit that will light both bulbs. Have them make the lights as bright as possible.
3. Have students draw circuit diagrams for the circuit created.

Experiment 2
1. Give students a hidden circuit box and a circuit tester. Have them use the circuit tester to find out how the wires are connected inside the box. Have them draw their predictions of the ways the wires are connected.

Post-Unit Assessments / **91**

APPENDIX A

Experiment 3

1. Make a more challenging hidden circuit box—one that has only two paper fasteners on the outside. Inside, make a connection between the two paper fasteners. The connection can be either a wire, a battery/battery holder, or a bulb/bulb socket connected to two of the paper fasteners.

   Use this box (Figure A-1) with more advanced students. It can be quite challenging.

Figure A-1

A challenging hidden circuit box

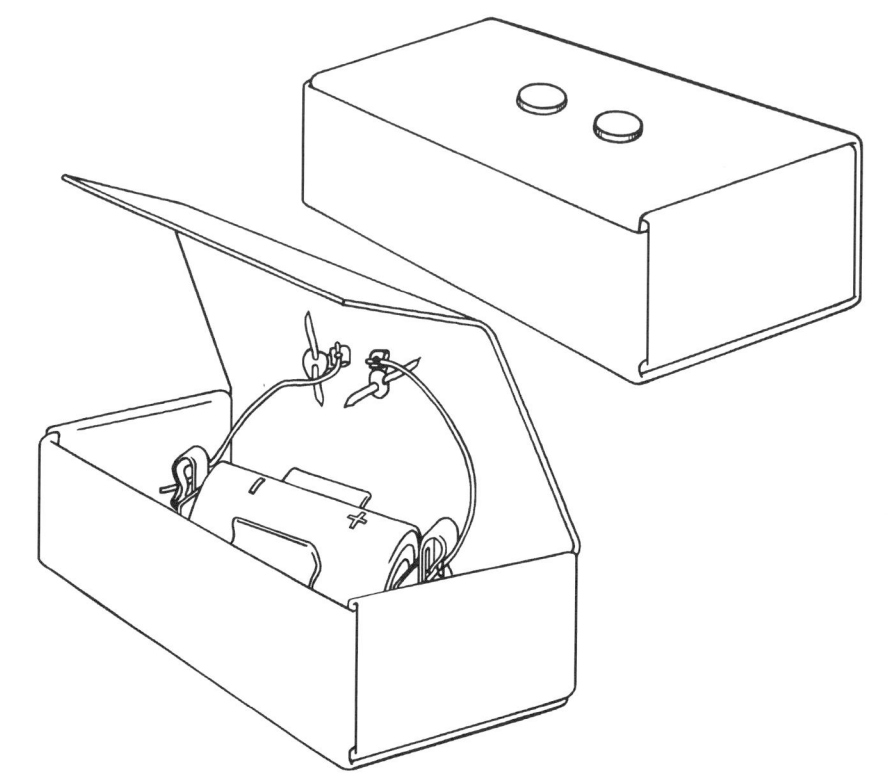

2. Tell students that this is a more challenging problem. Tell them that inside the box there may be a battery, a bulb, or a wire connecting the two exposed terminals. As they work on solving this problem, ask them to keep track of the tests they do and what they learn from the results.

3. The student's written record of the tests and the thinking that they have done provide the evaluation record. If a student can devise several tests and make valid conclusions from the results, then the student has learned a great deal, whether or not he or she is able to correctly determine the contents.

ASSESSMENT 5   **A Paper-and-Pencil Test for Students**

Materials   *For each student*

         Copies of **Evaluation Activity Sheet** (see pg. 93)

Post-Unit Assessment

Evaluation Activity Sheet

NAME: _____

DATE: _____

Find out how much you know about electricity by answering these questions.

1. Here is a picture of a circuit.

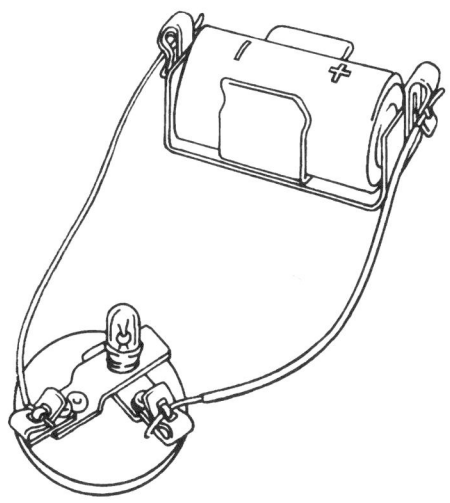

Using the symbols for the battery, the bulb, and the wire listed below, draw a circuit diagram for this circuit.

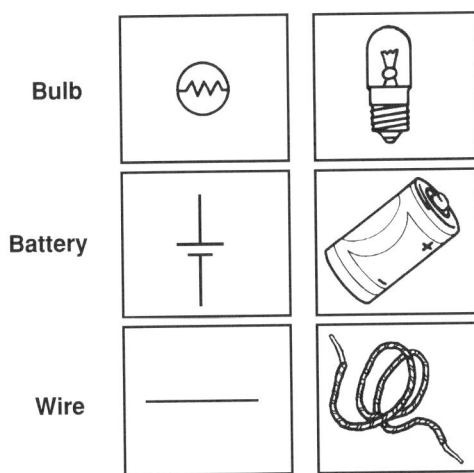

Post-Unit Assessment

2. Imagine you are walking down the street and you see a loose wire from a utility pole lying on the ground, with one end still attached to the pole. What are two safe things you might do?

3. You have just put a circuit together. After you have completed the job, the light does not go on as you expected it to. What are two things you would do to try to figure out why it did not go on?

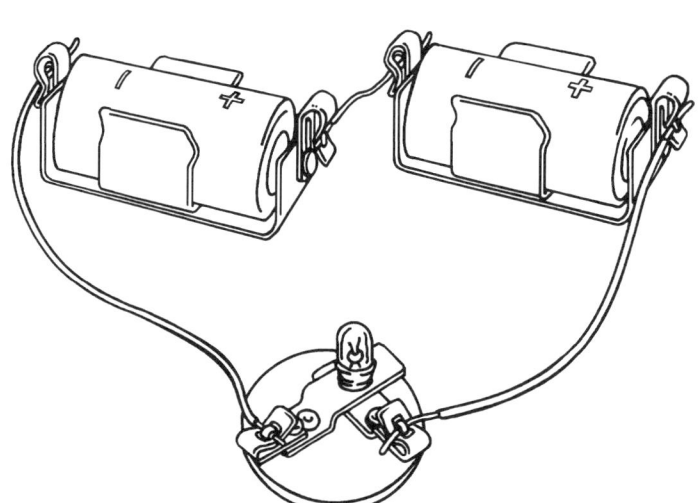

4. Here are some electrical devices connected by wires and a switch. The bulb is on now. What do you think will happen when the switch is closed?

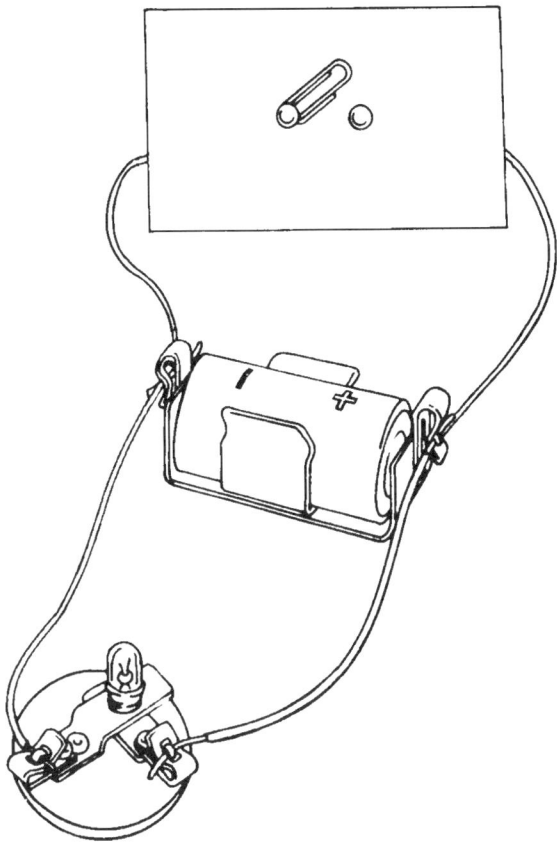

5. What do you think you learned about electricity this year?

6. What questions do you have about electricity now?

APPENDIX B

Teacher's Record Chart of Student Progress

Teacher's Record Chart of Student Progress for *Electric Circuits*

		Student
Products	Lesson 2: Drawing of lightbulb	
	Lesson 2: Drawings of flashlight bulb lighting	
	Lesson 3: Activity Sheet 1	
	Lesson 4: Drawing of household bulb	
	Lesson 5: Activity Sheet 2	
	Lesson 6: Activity Sheet 3	
	Lesson 7: List of conductors and insulators	
	Lesson 9: Drawings of hidden circuit boxes	
	Lesson 10: Drawing of circuit using symbols	
	Lesson 11: Drawing of series and of parallel circuit	
	Lesson 12: Drawing of circuit with a switch	
	Lesson 13: A flashlight	
	Lesson 13: Drawing of the circuit of their flashlight	
	Lesson 14: Drawing of circuit with diode	
	Lesson 15: Drawing of wiring plan for house	
	Lesson 16: A successfully wired house	
	Lesson 16: Drawing of house as wired	
Specific Skills	Can light flashlight bulb	
	Can draw inside of household bulb	
	Can demonstrate successfully the places to touch the bulb to light it. (Activity Sheet 3)	
	Can use the circuit tester and can locate circuit problems	
	Can use circuit tester to identify conductors and insulators	
	Can use circuit tester to solve hidden circuit boxes	
	Can use circuit diagram symbols to draw a circuit	
	Can predict the bulb brightness in series vs parallel circuits	
	Can create series and parallel circuits as needed	
	Can construct a switch and use it to turn a circuit on and off	
	Can construct a flashlight	
	Can successfully draw the wiring of the flashlight	
	Can put the diode in a circuit and demonstrate its one-way characteristic	
	Can wire a house so rooms are lighted and can be switched on and off	
	Can successfully draw the wiring of the house	
General Skills	Follows directions	
	Records observations with drawings or words	
	Works cooperatively	
	Contributes to discussions	

APPENDIX C

Using the Cutting and Stripping Tool

The wire that you use may be covered by a plastic insulating sheath. You will need to cut and strip this sheath from wire before you begin Lesson 1. To do that, strip away the cover with a wire stripping tool, two of which are shown in Figure C-1. (A small knife or wire cutter also can be used.)

Figure C-1

Wire stripping tools

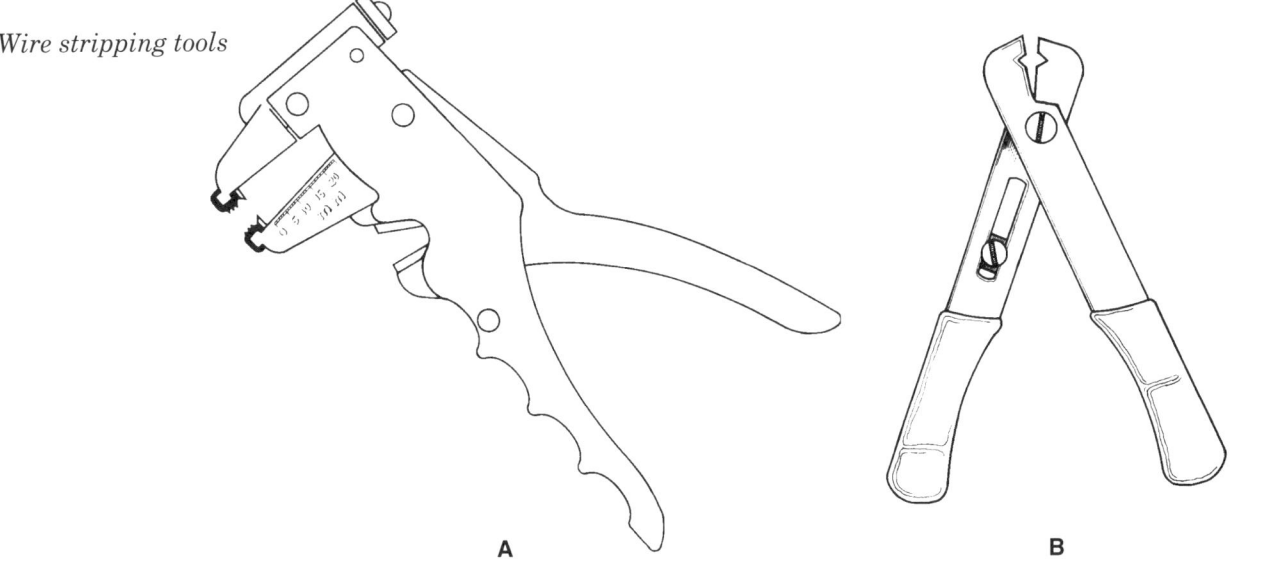

A B

Use a tool to cut the wire. Figure C-2 illustrates how to use the stripper shown in Figure C-1A. Figure C-3 shows how to use the cutter shown in Figure C-1B.

APPENDIX C

Figure C-2

Using a wire stripping tool

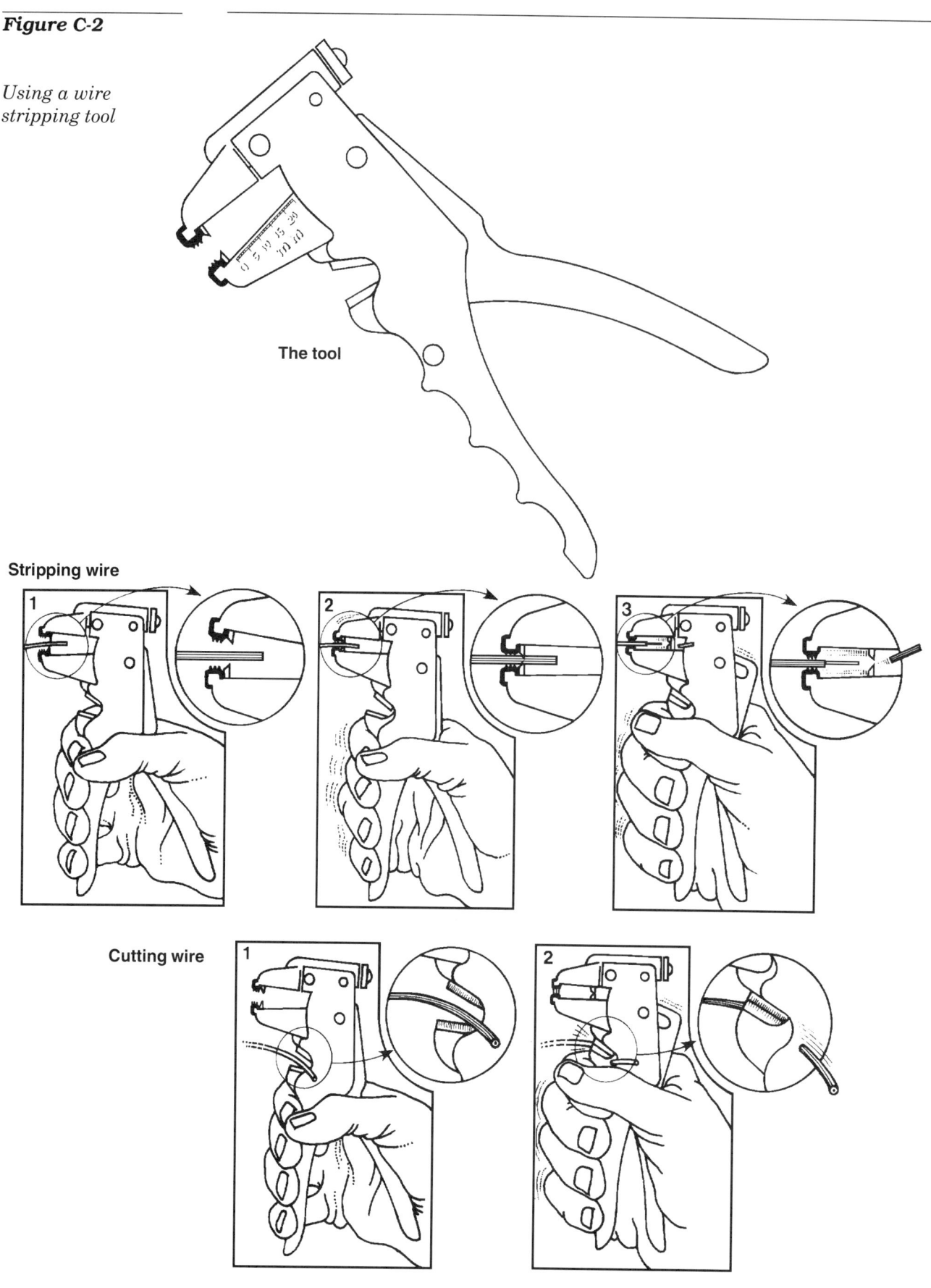

APPENDIX C

Figure C-3

Using a wire cutter

Cutting wire

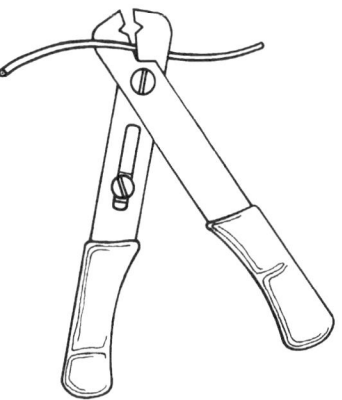

Make a small opening here...

...by adjusting this screw.

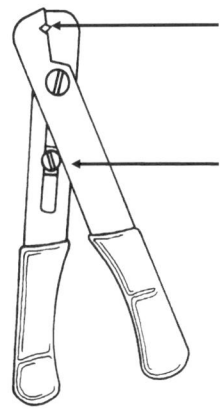

Put the wire in the opening and twist the cutters.

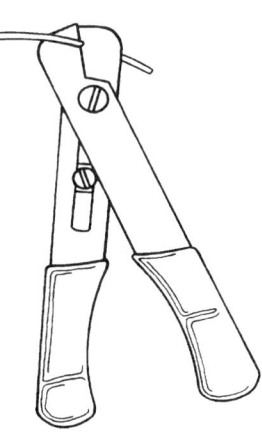

Pull the insulation off with the cutters.

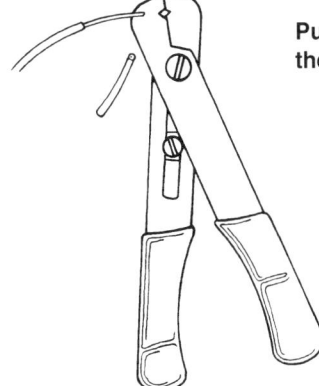

Using the Cutting and Stripping Tool / **101**

APPENDIX D

Removing the Base from a Light Bulb

The following instructions on how to remove the base from a light bulb will help you prepare for Lesson 4.

1. Take one of the household bulbs and carefully remove the base. The procedure, with drawings, is shown in Figure D-1. Use the small wire cutters and the needle-nose pliers. When the base is removed, you can see where the wires pass through the glass.

2. Save the metal parts of the base to show to the students.

3. Similarly, remove the base from one of the small bulbs. The technique for doing this is shown in Figure D-2.

APPENDIX D

Figure D-1

Removing the base from a household bulb

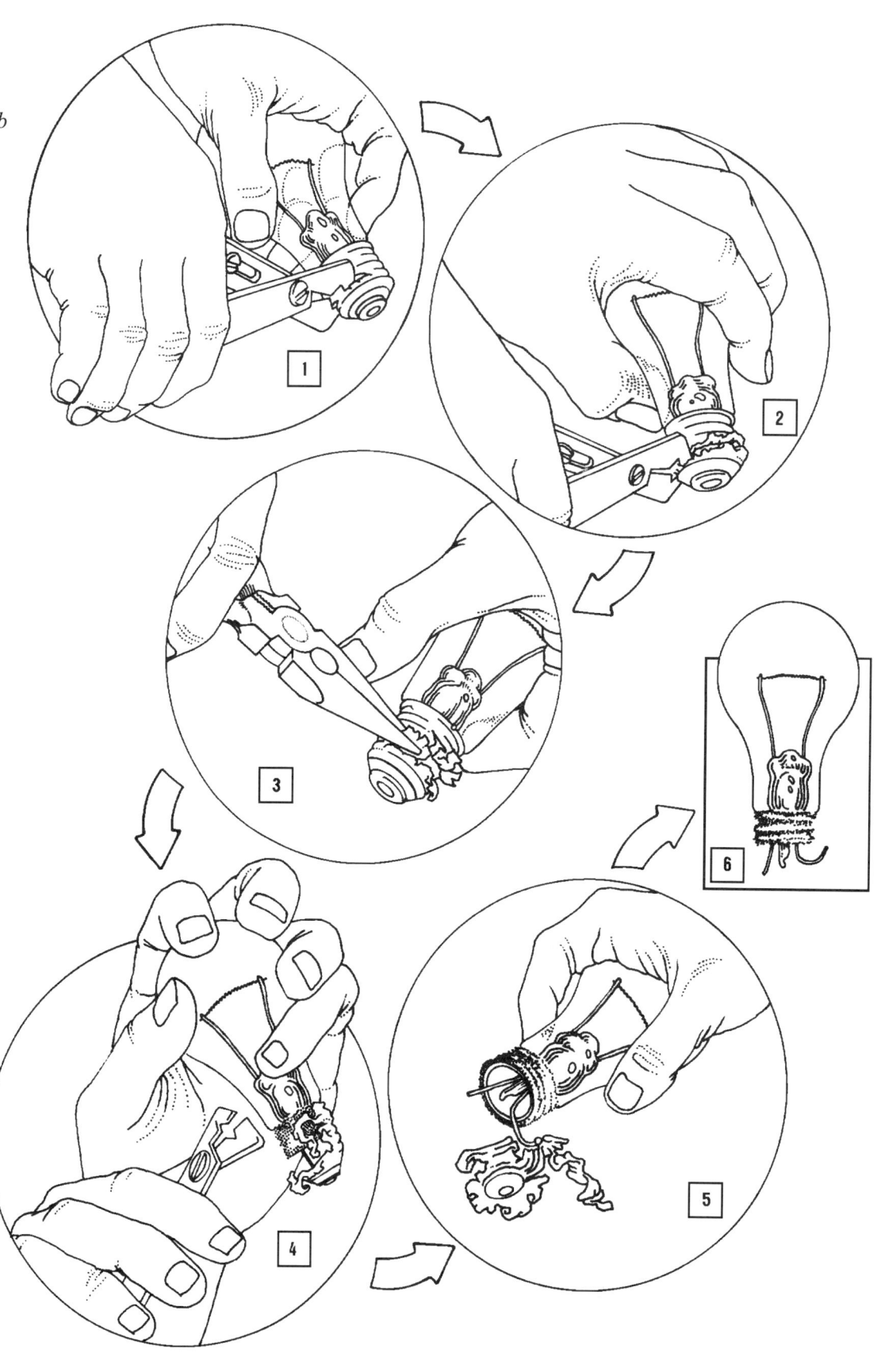

104 / Removing the Base from a Light Bulb

Figure D-2

Removing the base from the small bulb

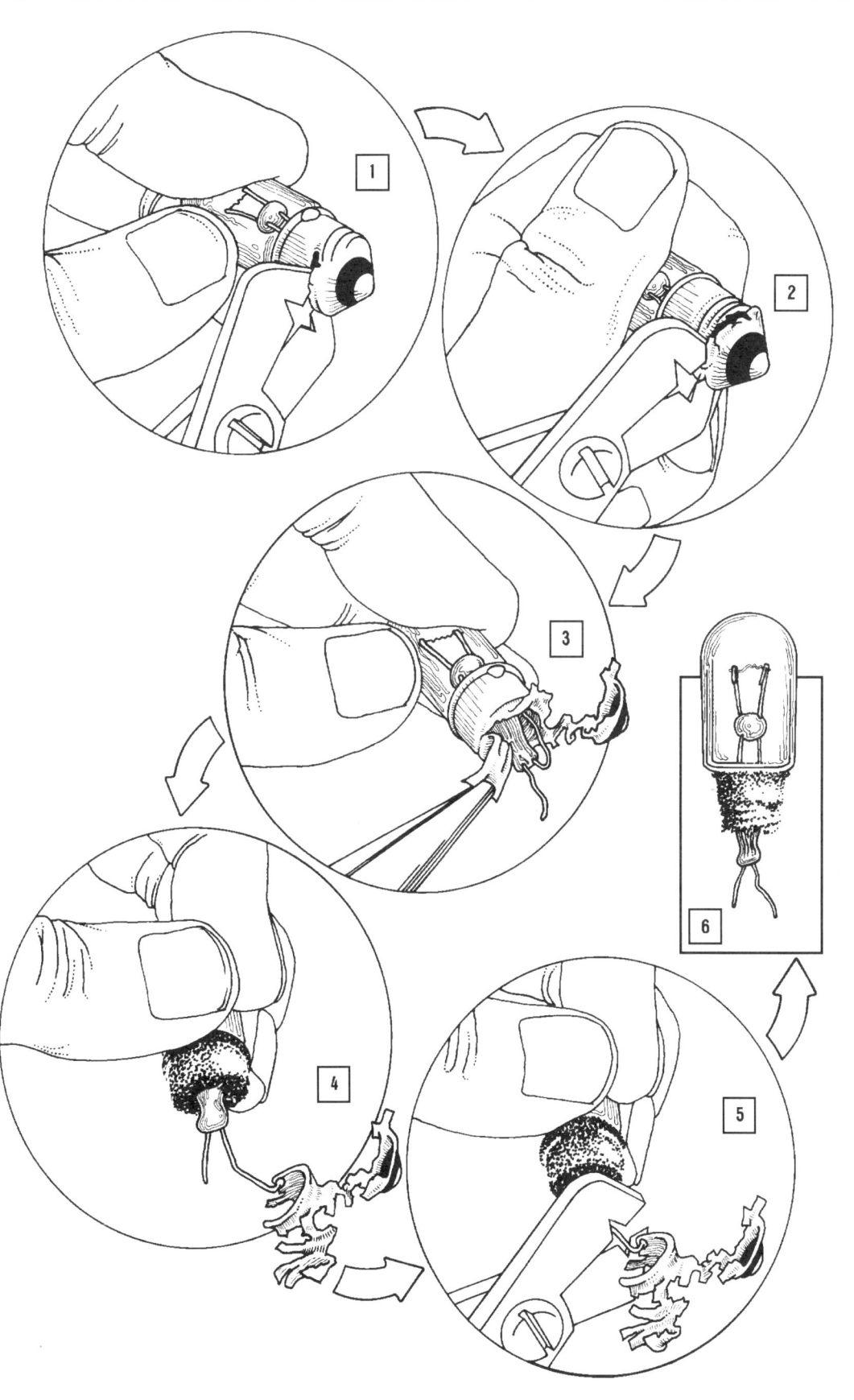

Removing the Base from a Light Bulb / **105**

APPENDIX E

Making and Troubleshooting Hidden Circuit Boxes

The following instructions are for making the hidden circuit boxes needed for Lesson 9. You can make the boxes yourself (including the one you will need for demonstration purposes). Or you can work with a few student helpers or parent volunteers. Make one box for every two students.

Figure E-1 shows a hidden circuit box.

Figure E-1

Hidden circuit box

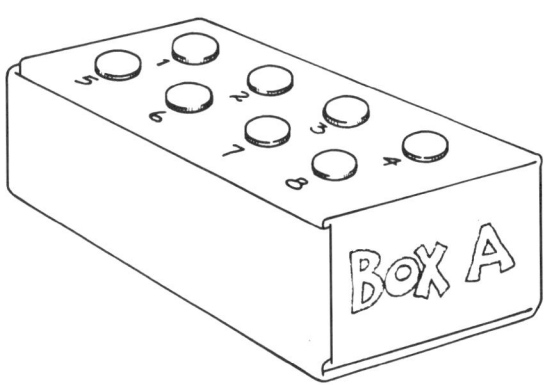

Materials

For each hidden circuit box (or card)
- 1 box (or two 5" x 8" index cards and tape)
- 8 No. 3 brass paper fasteners
- 8 brass paper fastener washers
- 8 Fahnestock clips
- 2 6-inch pieces of wire, stripped bare 1 inch at each end
- ¾" masking tape

Procedure

1. Using a screwdriver, punch two rows of four holes each on the lid. Place a brass paper fastener, with the head on top, in each of these slots.

    (Two 5" x 8" index cards can be used as an alternative to the box. One card is used like the lid of the box. The second card is taped as a backing to hide the wires.)

APPENDIX E

2. As you put in each paper fastener, hold the top with your finger. On the bottom, put on a Fahnestock clip and a brass paper fastener washer. Open the paper fastener ends, and push them apart firmly. Taping the paper fasteners into position on the underside of the lid helps eliminate the potential for their making contact with each other. Wires can be attached to the Fahnestock clips (see Figure E-2).

Figure E-2

Making the hidden circuit box

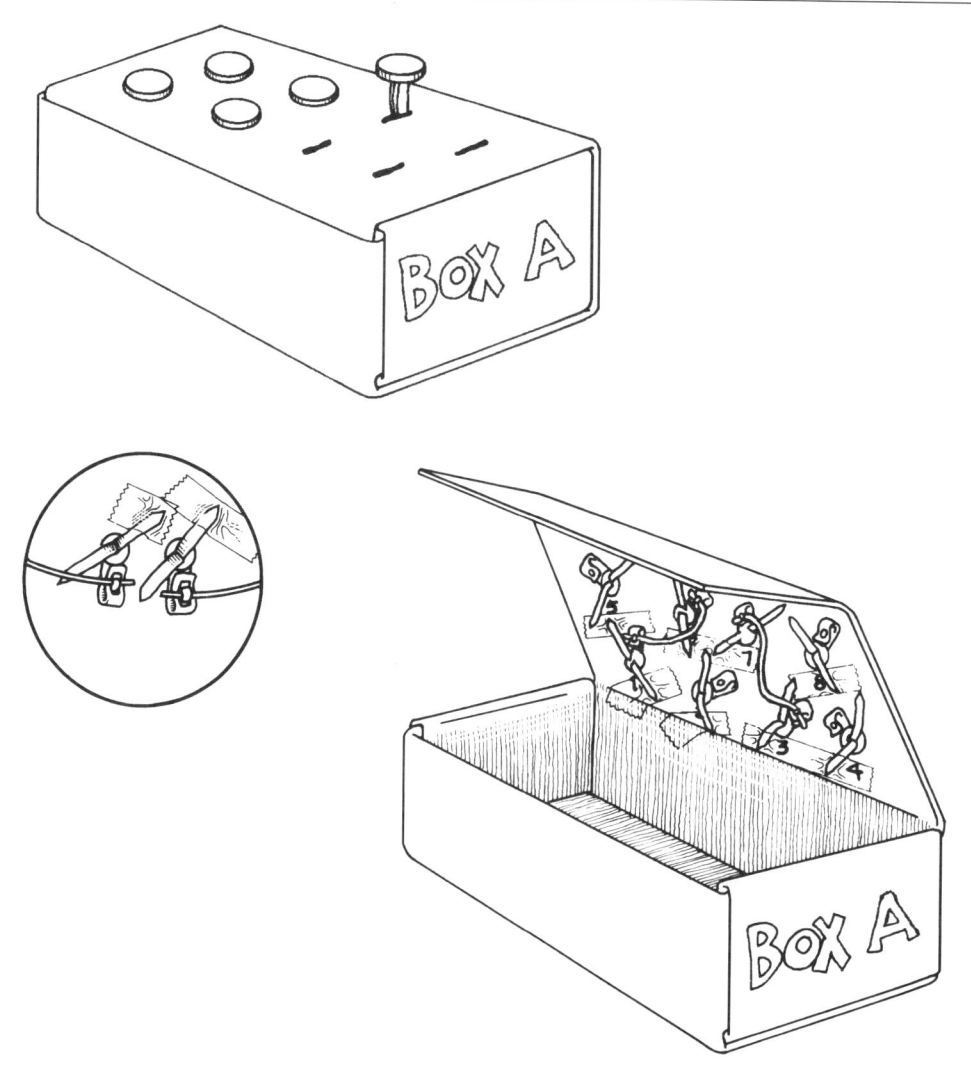

3. Wire each box differently.
4. Number the paper fasteners and give a letter designation for each box.
5. The Fahnestock clips make it easy to rewire each box to create a different circuit configuration.

Note: Sometimes the circuit boxes do not work because there is poor electric contact. Dirt, grease, or corrosion on the wire or the paper fasteners can be responsible for this. Wiping the surfaces with a piece of fine sandpaper or a clean paper towel usually will restore good electrical contact. Check also the wire connections on the Fahnestock clips, and make sure the ends of the paper fasteners are spread firmly.

APPENDIX F

Background for Lesson 11: Series and Parallel Circuits

There are two kinds of circuits: series circuits and parallel circuits. In a series circuit, electricity has only one path to travel, from one point on the circuit through the wires, batteries, and bulbs and then back to the beginning. In a parallel circuit, there is more than one path for the electricity to follow.

Keep in mind that batteries can be arranged in series or in parallel, and bulbs and other devices also can be connected in series or in parallel.

To understand what this means, imagine you are a very small person and are standing on a wire. You can walk in only one direction. If you can walk along the wire and never have a choice about which way to turn, the circuit is entirely in series. On the other hand, if there are "forks in the road" where you have two directions to turn, then you have stumbled onto a circuit that is in parallel.

Think about the wiring in your house. When you have two different lamps plugged into one receptacle, one lamp being turned off or one bulb burning out does not make the other lamp go off. This shows that these two lamps are in a parallel circuit arrangement.

On the other hand, think about the old Christmas tree ornaments, where one light burning out meant that the whole string went out. Those lights were in a series arrangement.

Some series circuits are shown in Figures F-1, F-2, and F-3. Some parallel circuits are shown in Figures F-4, F-5 and F-6.

Make these circuits and you will note that the bulb is brighter in Figure F-1, where the batteries are in series, than in Figure F-4, where they are in parallel.

By contrast, compare the brightness of the bulbs in Figure F-2 with that in F-5. Notice that the two bulbs connected in parallel (Figure F-5) are brighter than the ones connected in series (Figure F-2).

APPENDIX F

Figure F-1

Two batteries in series, one bulb

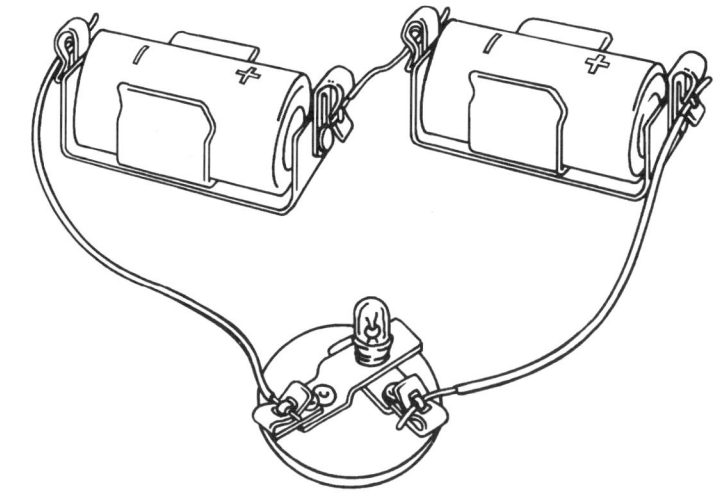

Figure F-2

Two bulbs in series, one battery

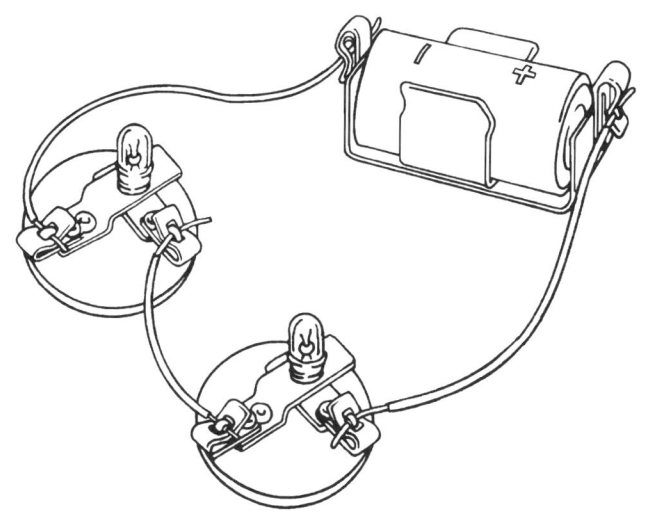

Figure F-3

Two bulbs in series, three batteries in series

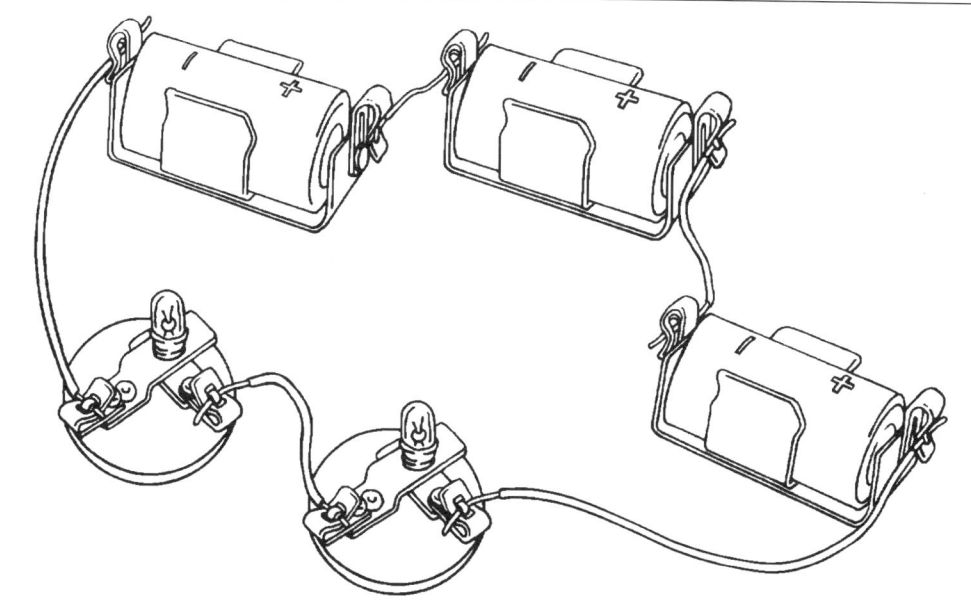

Figure F-4

Two batteries in parallel, one bulb

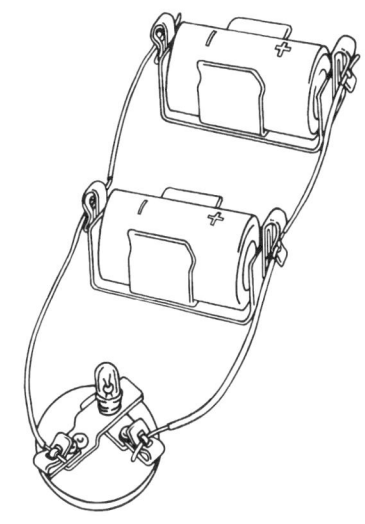

Figure F-5

Two bulbs in parallel, one battery

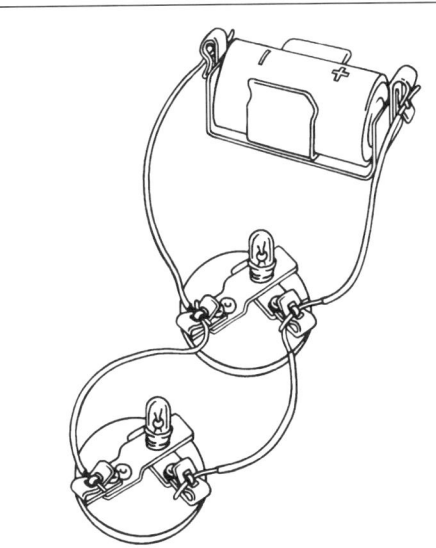

Figure F-6

Three batteries in series, three bulbs in parallel

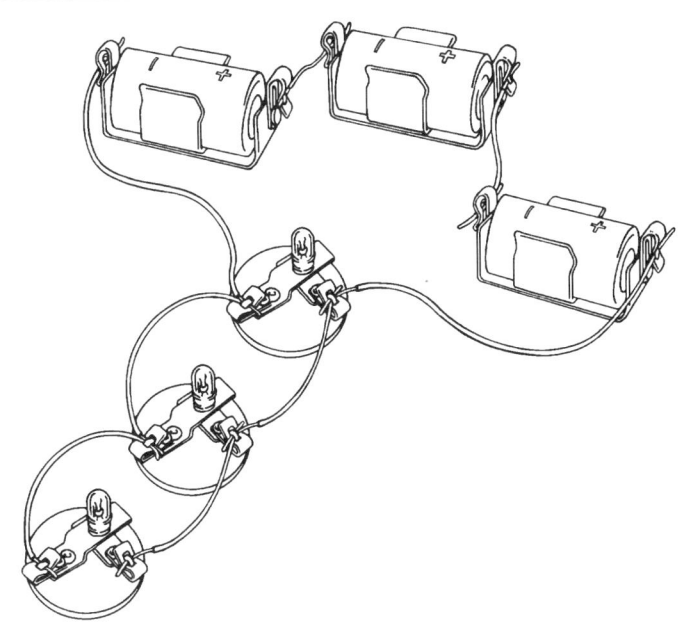

Background for Lesson 11: Series and Parallel Circuits / **111**

APPENDIX G

Bibliography

Resources for Teachers

Dishon, Dee, and O'Leary, Pat Wilson. *A Guidebook for Cooperative Learning: Techniques for Creating More Effective Schools*. Holmes Beach, Florida: Learning Publications, Inc.,

> A practical guide for teachers who are embarking on the implementation of cooperative learning techniques in the classroom.

James, Portia P. *The Real McCoy: African Invention and Innovation, 1619-1930*. Washington, D.C.: Smithsonian Institution Press, 1989.

> *The Real McCoy* tells of the creative spirit of black men and women from the time of the earliest settlements through slavery and emancipation to modern times. The electrical inventions of Granville Woods and Lewis H. Latimer are dealt with in the last pages of the book. While the individual accounts are brief, the photographs and drawings provide a unique historical context for their work.

Johnson, David W., Johnson, Roger T., and Holubec, Edythe Johnson. *Circles Of Learning*. Alexandria, Virginia: Association for Supervision and Curriculum Development, 1984.

> This excellent book presents the case for cooperative learning in a concise and readable form. It reviews the research, outlines implementation strategies, provides definition to the skills needed by students to work cooperatively, and answer many questions.

Resources for Students

Beasant, Pam. *Introduction to Electronics*. Tulsa: Educational Developmental Corp., 1985.

> This is a simple, but greatly detailed introduction to electronics for beginners. It is very amply illustrated and targeted at the younger reader. The interested student will be able to use it to make use of much of what is being learned in this unit.

APPENDIX G

Chapman, Philip. *Electricity (The Young Scientist Book of).* Tulsa: Educational Developmental Corp., 1976.

> This book gives some interesting details about how electricity is generated, and a brief account of batteries. It shows a detailed cutaway view of the inside of a D-cell. The book goes well beyond the scope of this unit, and may answer, briefly and concisely, many of the questions that come up.

Cosner, Shaaron. *The Light Bulb.* New York: Walker and Company, 1984.

> This is a serious book for young readers who are interested in the story of Edison's invention of the light bulb. It provides enough of the details of his struggles and the gritty work involved to give the budding young inventor a realistic view of the tasks that are ahead.

Davidson, Margaret. *The Story of Benjamin Franklin, Amazing American.* New York: Bantam Doubleday Dell, 1988.

> This is an excellent, lively biography. It captures Franklin's curiosity and inventiveness while maintaining the very human dimensions of the man. This book is worth reading both for the science and for the history.

Fritz, Jean. *What's the Big Idea, Ben Franklin.* New York: Putnam Publishing Group, 1976.

> This charming biography is excellent for the history and for the down-to-earth qualities it captures of Franklin's life. His systematic problem solving, leading to many useful inventions, is very well presented. This book is worth reading both for the science and for the literary value.

Haber, Lewis. *Black Pioneers of Science and Invention.* New York: Harcourt, Brace, Jovanovich, Inc., 1970.

> This book traces the lives of black scientists and inventors who have made significant contributions in the various fields of science and industry. Of particular interest to the study of electricity and its applications are the lives of Granville T. Woods and Lewis H. Latimer.

James, Portia P. *The Real McCoy: African Invention and Innovation, 1619-1930.* Washington, D.C.: Smithsonian Institution Press, 1989.

> *The Real McCoy* tells of the creative spirit of black men and women from the time of the earliest settlements through slavery and emancipation to modern times. The electrical inventions of Granville Woods and Lewis H. Latimer are dealt with in the last pages of the book. While the individual accounts are brief, the photographs and drawings provide a unique historical context for their work.

Math, Irwin. *Wires and Watts: Understanding and Using Electricity.* New York: Charles Scribner's Sons, 1981.

> The book uses experiments and projects that produce actual working models to present the fundamentals of electricity. It goes considerably beyond the content of *Electric Circuits*, introducing the use of simple equations to express Ohms law, detailing the working of a D-cell battery, and explaining a variety of instruments and devices. This

book would be of interest to the most advanced and ambitious students, who could use it with an adult.

Quackenbush, Robert. *Quick, Annie, Give Me A Catchy Line, A Story of Samuel F. B. Morse*. New York: Prentice-Hall, Inc., 1983.

This is a whimsical account of Samuel Morse's life and his invention of the telegraph. It captures some interesting parts of an unusual history.

Sabin, Louis. *Thomas Alva Edison, Young Inventor*. Mahwah, New Jersey: Troll Associates, 1983.

A short, readable paperback that reveals some of the extraordinary characteristics of the young Edison. It does not give any details of his work leading to the invention of the light bulb, but it does offer a fetching view of his industriousness and inventiveness as a young man.

Materials Reorder Information

During the course of hands-on science activities, some of the materials are used up. The consumable materials from each Science and Technology for Children unit can be reordered as a unit refurbishment set. In addition, a unit's components can be ordered separately.

For information on refurbishing *Electric Circuits* or purchasing additional components, please call Carolina Biological Supply Company at **800-334-5551** and ask for an STC Customer Service Representative.

National Science Resources Center Advisory Board

Chairman
Robert M. Fitch, Senior Vice President (retired), Research and Development, S. C. Johnson Wax, Racine, WI

Members
Russell Aiuto, Senior Project Officer, Council of Independent Colleges, Washington, DC
Marjory Baruch, Educational Consultant, Fayetteville, NY
Ann Bay, Director, Office of Elementary and Secondary Education, Smithsonian Institution, Washington, DC
DeAnna Banks Beane, Project Director, YouthALIVE, Association of Science-Technology Centers, Washington, DC
F. Peter Boer, Executive Vice President and Chief Technical Officer, W. R. Grace and Company, Boca Raton, FL
Douglas K. Carnahan, Vice President and General Manager, Measurement Systems Organization, Hewlett-Packard Company, Boise, ID
Fred P. Corson, Vice President and Director, Research and Development, The Dow Chemical Company, Midland, MI
Goéry Delacôte, Executive Director, The Exploratorium, San Francisco, CA
JoAnn E. DeMaria, Elementary School Teacher, Hutchison Elementary School, Herndon, VA
Hubert M. Dyasi, Director, The Workshop Center, City College School of Education (The City University of New York), New York, NY
Bernard S. Finn, Curator, Division of Electricity and Modern Physics, National Museum of American History, Smithsonian Institution, Washington, DC
Gerald D. Fischbach, Department of Neurobiology, Harvard Medical School, Boston, MA
Samuel H. Fuller, Vice President of Corporate Research, Digital Equipment Corporation, Littleton, MA
Ana M. Guzmán, Program Director, Alliances for Minority Participation, Texas A & M University, College Station, TX
Robert M. Hazen, Staff Scientist, Carnegie Institution of Washington, Washington, DC
Norbert S. Hill, Jr., Executive Director, American Indian Science and Engineering Society, Boulder, CO
Manert Kennedy, Executive Director, Colorado Alliance for Science, University of Colorado, Boulder, CO
John W. Layman, Professor of Education and Physics, and Director, Science Teaching Center, University of Maryland, College Park, MD
Sarah A. Lindsey, Science Coordinator, Midland Public Schools, Midland, MI
Thomas E. Lovejoy, Counselor for Biodiversity and Environmental Affairs, Smithsonian Institution, Washington, DC
Lynn Margulis, Professor of Biology, Department of Botany, University of Massachusetts, Amherst, MA
Shirley M. McBay, President, Quality Education for Minorities Network, Washington, DC
John A. Moore, Professor Emeritus, Department of Biology, University of California, Riverside, CA
Philip Needleman, Corporate Vice President, Research and Development, and Chief Scientist, Monsanto Company, St. Louis, MO
Carlo Parravano, Director, Merck Institute for Science Education, Rahway, NJ
Ruth O. Selig, Executive Assistant to the Acting Provost, Smithsonian Institution, Washington, DC
Maxine F. Singer, President, Carnegie Institution of Washington, Washington, DC
Paul H. Williams, Director, Center for Biology Education, and Professor, Department of Plant Pathology, University of Wisconsin, Madison, WI
Karen L. Worth, Faculty, Wheelock College, and Senior Associate, Urban Elementary Science Project, Education Development Center, Newton, MA

Ex Officio Members
E. William Colglazier, Executive Officer, National Academy of Sciences, Washington, DC
James C. Early, Assistant Provost for Educational and Cultural Programs, Smithsonian Institution, Washington, DC